Dragon SONGS

Dragon SONGS

Love and Adventure
Among Crocodiles, Alligators, and
Other Dinosaur Relations

VLADIMIR
DINETS

Arcade Publishing • New York

First Edition

Arcade Publishing books may be purchased in bulk at special discounts for sales promotion, corporate gifts, fund-raising, or educational purposes. Special editions can also be created to specifications. For details, contact the Special Sales Department, Arcade Publishing, 307 West 36th Street, 11th Floor, New York, NY 10018 or arcade@skyhorsepublishing.com.

Arcade Publishing® is a registered trademark of Skyhorse Publishing, Inc.®, a Delaware corporation.

Some of the photographs in the insert were coauthored by Alex Bernstein, Sarit Reizin, and Anastasiia Tsvietkova.

Visit our website at www.arcadepub.com.
Visit the author's website at dinets.info

10 9 8 7 6 5 4 3 2 1

Library of Congress Cataloging-in-Publication Data

Dinets, Vladimir.
Dragon songs : love and adventure among crocodiles, alligators, and other dinosaur relations / Vladimir Dinets. — First edition.
pages cm
ISBN 978-1-61145-893-0 (alkaline paper)
1. Crocodiles—Behavior. 2. Alligators—Behavior. 3. Animal communication.
4. Sexual behavior in animals. 5. Crocodiles—Ecology. 6. Alligators—Ecology. 7. Dinets, Vladimir—Travel. I. Title.
QL666.C925D54 2013
597.98'2—dc23

2013028034

Printed in the United States of America

To Steven Green

Contents

List of Photographs

List of Maps and Charts

List of Maps and Charts

Acknowledgments

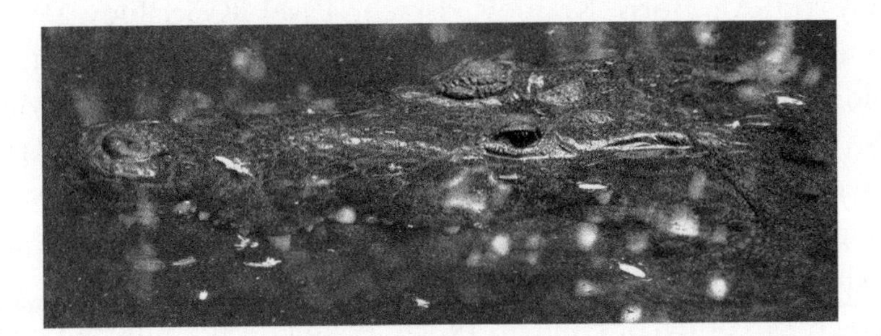

A journey is best measured in friends rather than miles.
—Buddha

MANY PEOPLE HAVE MADE THIS BOOK POSSIBLE, often going beyond what they were asked for, sharing their time, knowledge, experience, transportation, food, water, homes, and campfires, issuing research permits, allowing access to private lands, tribal territories, and research facilities. I can name only a selected few here, but am grateful to all of them.

My thanks to Eisha Alibasha (Ethiopia), Jamal Al-Jawari (Egypt), Chief Asanka (Burundi), Balaqa Ashoba (Ethiopia), Mark Barrett (USA), Ben Bejtel (Namibia), Alex Bernstein (Kenya), Gaurav Bhatnagor (India), Tomas Blohm (Venezuela), Jesekiel Bomba (Uganda), John Brueggen (USA), Jaco Burger (Namibia), Gordon Burghardt (USA), Rick Cameron (Indonesia), Zilca Maria da Silva Campos (Brazil), Mariela Cardenas (Ecuador), Ada Castillo (Peru), Joseph Chadou (DRC), Michael Cherkis (USA), Jennifer Ciaccio (USA), John Clenn (Jamaica), Tommy Collard (Namibia), Steven Conners (USA), Steven Counsel (Mozambique), Paul van Damme (Brazil), Alla Dinets (Russia), Mitch Eaton (USA), Beniamin Eligulashvili (Israel), Tamrat Fahid (Ethiopia), Carla Ferreira (South Africa), Bereket Gebretsadik (Ethiopia), Carl Gerhard (USA), Steven Green (USA), John Hall (USA), Casey Handmer (Australia), Dabeer Hasan (India), Paolo Hatzfeld (Brazil), Steve Irvin (Australia), Tim Isle (USA), Tazz Jacobs (South Africa), Zhao Xin Jiang (China),

Alan Karlon (USA), Michael L. Kipkeu (Kenya), Avid Kledzik (USA), Ofer Kobi (Israel), Aaron Kortenhoven (USA), Michael Kosoy (USA), Boris Krasnov (Israel), Pavel Kvartalnov (Russia), Jeronimo Dominguez Laso (Mexico), Mashera Leboma (Tanzania), Pablo Lopez (Ecuador), Raul Marago (Bolivia), Thaddeus McRae (USA), Diana, Edward, and Melanie McTurk (Guyana), Krishna Kumar Mishra (India), Liz von Muggenthaler (USA), Mzwandile Mjadu (South Africa), Pedro Maria Montero (Colombia), Soham Mukherjee (India), Mary N. (Madagascar), Darren Naish (UK), Carmen N. (Bolivia), François Ndele (Cameroon), Sunday Nelenge (Namibia), Jeremiah Newman (DRC), James C. Nifong (USA), David Oudjani (France), Maria Oleneva (Mexico), Serverio Pachac (Ecuador), Richard Paper (Madagascar), Asy Patrysheva (Russia), Elena and Victor Pavlov (USA), Søe Pedersen (Denmark), Dietmar Posch (South Africa), Pradeep (India), Alfonso Llobet Querejazu (Bolivia), Devis Rachmawan (Indonesia), Eulalie Razoanantenaina (Madagascar), Patty and Allen Register (USA), Wang Renping (China), Sarit Reizin (Kenya), Mark Robinson (South Africa), Damian Rumiz (Brazil), Jorge Sanchez (Guatemala), William Searcy (USA), Genzap Sechen (Ethiopia), Mejangi Sfatau (Uganda), Kabir Sharma (India), Lorraine Shaunnessy (USA), Gopi Shindar (India), Victor M. Siamudaala (Zambia), Kesi Sinclair (USA), Allan Smale (South Africa), Ralf Sommerland (Germany), Jievu Taljard (South Africa), Megan Taplin-Brandfield (South Africa), Peter Taylor (Guyana), Jean-Claude Toutondele-Louzolo (Congo), Asteraye Tsigie-Tesfahunegn (Ethiopia), John Thorbjarnarson (USA), Steve Thorley (Zimbabwe), Richard Tokarz (USA), Kathryn Tosney (USA), Anastasiia Tsvietkova (USA), Ramon Vaquero (Venezuela), Mike Veitch (Indonesia), Eric Villaume (Gabon), Kent Vliet (USA), Roland Vorwerk (USA), Keith Waddington (USA), Nikhil and Romulus Whitaker (India), Christy Wolovich (USA), Xinhuan Wang (China), Utai Youngprapakorn (Thailand), Tesfaye Zewdie (Ethiopia), Benyamin Zuwadi (Indonesia), and Stella !Nomxas (Namibia); also the personnel of the Nad-Lembeh and Papua Diving dive centers (Indonesia), Napo Wildlife Center (Ecuador), and all protected natural areas, publicly or privately owned, where I've conducted my research.

Dragon SONGS

Prologue

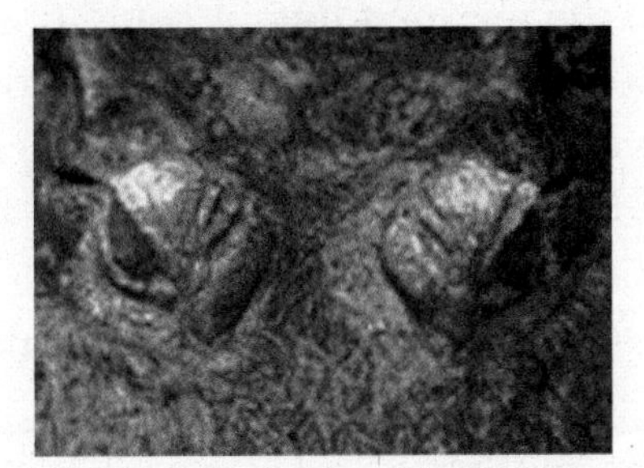

A true nomad is never lost.
—Evenk proverb

IT WAS MY LUCKY DAY. At eight in the morning I discovered that I had tropical malaria, and I was really happy about it.

I first felt it two days earlier, while hiking down from the summit of Mount Kenya. I had to leave my backpack in trailside bushes before the ascent, and, of course, it got stolen. Things left unattended in Africa tend to disappear fast, even at ten thousand feet above sea level. No longer having a sleeping bag, I had to walk all night to get to the highway. The next day I felt really weak, and thought I was just overtired. But the fatigue was getting worse; I developed a fever and was sweating heavily. It reminded me of pneumonia, and pneumonia would require daily penicillin shots, which would be difficult to arrange while bush-camping.

So I was very glad to have my first typical malaria attack. It lasted only a few minutes, while I was sitting in a café trying to force down some manioc puree. It's amazing that malaria parasites, these rogue descendants of innocent marine algae, living and multiplying inside your red blood cells, can synchronize their mass emergence so tightly. I hopped on a bus to Nairobi, went to a hospital, let them do a blood test, bought some Coartem, took the first two pills, and got perfectly well within three hours. And, of course, I asked the doctor who was checking the test results to give me the microscopic slide with my blood sample as a souvenir. I still have it, and show it to visiting friends.

You can clearly see little purple *Plasmodium falciparum* parasites inside some red cells—pretty cool, I think.

I checked into the cheapest hotel I could find. Traveling in Africa is expensive, and I was near the end of a four-month journey. I ended up in a part of town where the hotel owner felt obliged to give me an assault rifle–armed guard every time I walked to the grocery store across the street that doubled as an Internet café. On my third guarded trip the connection was so good that I finally managed to check my email in less than forty minutes.

First, I sent my mother a coded email: "ok vova." *Vova* is short for Vladimir in Russian. *OK* meant "Everything is going according to the plan; I'm perfectly well and traveling in wonderful places." My mother and I had developed this code in pre-Internet days for use in telegrams, but it also came handy in Africa where online connections tended to get lost unexpectedly.

After making sure my mother knew I was alive, I went on to read my emails—and there was one informing me that I'd been admitted to the University of Miami in Florida as a graduate student.

This was the lucky break I'd been waiting for. It meant that I would never have to work again in my life. Professional zoology does not count as work in my book. It's not something you'd do only for pay. It can be more demanding—physically and mentally—than almost any other job, but it's so much fun for someone like me that it's worth the trouble. I'd been doing it as a volunteer or somewhat-paid field technician for a few years, but wages in zoology are much lower than in other sciences, and you pretty much need a PhD to survive without other income. So that's what I was headed for—a PhD in zoology—now that I had my acceptance from the university in hand.

I had to get back to the States as fast as I could. I had an open-date ticket that I'd bought from a shady Pakistani airline through an even shadier Turkish travel agent. I was supposed to fly to New York via Addis Ababa, Tripoli, and London. The first leg was by Ethiopian Airlines. But when I got to the airport, the girl at their counter informed me that Nairobi-Addis flights had been canceled due to street riots in Ethiopia. I was, of course, disappointed . . . and that's when this day proved to be a really lucky one.

"Hakuna matata" (everything's fine), said the girl. "My cousin is an officer; he runs armed convoys to Mogadishu. Maybe he can smuggle you into Somalia, and put you on our flight from there to Addis."

I thought she was joking, but she wasn't. She didn't have to help me, yet she didn't hesitate to call her cousin with a crazy request to take a foreigner on a military convoy across a closed border. I don't know what she told him in Swahili, but it took her less than three minutes to persuade him.

Since my backpack had been stolen, I had no luggage but what could fit in my pockets, so traveling became easy. I hitchhiked to the border and joined the convoy, which included two armored personnel carriers and a few old trucks, loaded with the stuff most wanted in civil war–ravaged Somalia: rap music tapes and cases of Coca-Cola. On their way back, those trucks would carry *qat* (a kind of mild narcotic popular in the Horn of Africa) for Somalis living in Kenya. I was allowed to ride on top of the lead carrier—the only place not engulfed in dust—but had to hide inside if there were people around. On the outskirts of Mogadishu, we were met by a tank that escorted us to a military compound. I wouldn't have minded a look at the city, but my hosts insisted on putting me on a cargo plane right away.

I got stuck for a few days in Addis Ababa: the airport was paralyzed by heavy rains in India. There are millions of Indians living in Africa, and almost all air traffic between their old and new worlds goes through Addis. On this day, most of it was grounded in India. At that time there were no ATMs in Ethiopia, and I had no cash left, but before I starved to death I managed to squeeze onto a plane to Washington DC, but it got hit by lightning on takeoff, but there wasn't much damage, but . . . Africa is not for people who don't like surprises.

All that time, covering mile after mile of dry savanna, freezing in over-air-conditioned airports, looking through plane windows at billowing tropical clouds, I felt very happy . . . and a bit sad.

Happy, because after such a long wait I could finally get back to doing things I liked and was good at. It had been eight years since, at the age of twenty-eight, I moved from decaying Russia to the

United States, but in all these years I couldn't get a good permanent job in zoology. At first I had to do mundane and low-paid things like logging or delivering pizza. I learned the hard way that in the world of free enterprise, the higher you climb, the less you have to work; the most difficult, demanding, and unpleasant occupations are also the ones on the low end of the pay scale. Then I managed to get semi-volunteer positions on various research projects. This work was interesting and enjoyable: studying whales and waterfowl in California, plankton in the Sargasso Sea, plague in prairie dogs on the Great Plains, and hantavirus in mice in New Mexico. But these projects weren't mine—I was just a hired hand. Now I finally had a chance to do my own research. I'd still be paid very little, but my work would be so interesting that I'd enjoy my life to the fullest anyway.

But I was a bit sad, too, because these temporary jobs could be easily abandoned, giving me long vacations in between. I liked travel more than anything since I was about ten. The former Soviet Union was a safe enough place for a kid to start hitchhiking around at about twelve. Over the years I learned to travel around the globe cheaply and easily (not always comfortably, but I didn't care), and could afford a trip to almost any part of the world after just a few months of work in a grocery store or a dot-com start-up. I enjoyed more freedom than most people could ever dream of. And now I would have to settle in a permanent home—something I always dreaded. My happy days of wandering freely were over.

Or so I thought.

Russians have a funny superstition. They half-seriously believe that the first day of a new year is like a preview of how the rest of it will go. If you spend January 1 happily drinking and dancing with your friends, you are OK until December 31. If instead you have a terrible migraine, lose your wallet, and your upstairs neighbors flood your apartment, expect 364 more days of trouble.

The first six days of this story were certainly an accurate depiction of what the whole six years would be like: intense, unpredictable, full of adventures and discoveries from beginning to end. And with more travel than I ever expected.

1
The Morning Chorus
Alligator mississippiensis

Study nature, not books.
—Louis Agassiz

I LANDED IN WASHINGTON DC, TOOK A BUS TO ALBUQUERQUE, where all my earthly possessions were waiting for me in a friend's garage, rented a truck, hooked my little Toyota Celica behind it, drove to Miami, and found a small apartment in Little Havana. Miami didn't look that different from Nairobi, except the heat was much more oppressive, the groves of downtown high-rises much uglier, the swamps contained more alligators than Africa had crocodiles, and people looked bloated compared to skinny Africans, as if half of them had some terrible tropical disease.

My arrival was perfectly timed: I had just enough time to settle in before two hurricanes hit our area. It would be really sad to miss such impressive storms.

My first semester in the university went by quickly. I enjoyed every minute of it. I had been interested in zoology to the point of obsession since I was a small kid but never had a chance to study it in an organized way. In the Soviet Union, where I grew up, the main universities were off-limits (unofficially, of course) for a person of Jewish origin like me, so I had to get my master's in a place more technology-oriented, studying "medical and biological equipment"—the closest I could get to what I really liked. By the time I graduated, the empire was falling apart and science was no longer something you could do for a living. So I became a freelance naturalist, writing guidebooks like *Mammals of Russia* and *Hitchhiking in South America*, leading bird-watching tours, filing environmental assessments, usually having two or three jobs at a time.

Now, at last, I could simply be a zoologist. But I was facing another problem: What animal to study? To me, all animals were interesting. Until now I could work with all of them in turn, switching from insects to whales, from snakes to mollusks, from plankton to fishes. It would be painful to become an expert on, say, hummingbirds and forget about all the other wonderful creatures. So I decided to specialize in studying animal behavior. This part of zoology is called ethology (from the Greek word *ethos*, "habit") in other countries, but for some reason this term is seldom used in the States. I would still be able to study whatever animal I liked, without having to do unpleasant things that many zoologists do: collect specimens, dissect living creatures, kill the very animals I liked enough to make them my vocation. Instead, I'd mostly observe animals doing things *they* like, and sometimes stage simple experiments to understand how and why they make their moves, choices, and decisions.

It was time to choose the subject of my PhD thesis research. Even after deciding that my study would be on animal behavior, I still had to pick a particular problem to work on. I had a few suggestions, ranging from studying petrel navigation near the magnetic poles to snow-tracking wolves in Tibet. But when I presented these ideas to Steven Green, my scientific advisor, he found them totally impractical and didn't hesitate to point out why. Steve is a brilliant scientist, and I enjoyed every moment of working with him, but he's not the kind of person who politely keeps silent when you make a mistake, and his style of teaching is a direct opposite of the make-you-feel-good approach popular in American schools. Finally, he got tired of shooting down one idea of mine after another and said:

"Why don't you look into alligator behavior? They do some interesting communication in spring. Garrick studied it, but that was almost thirty years ago. And you wouldn't have to travel too much; there're so many gators around here."

I didn't like the idea. As any biologist would, I found alligators—as well as crocodiles, caimans, and similar creatures, known together as crocodilians—fascinating from an evolutionary point of view. They are often called the last survivors from the Age of Reptiles (which lasted from about 250 to about 65 million years ago), living fossils, the closest thing to dinosaurs . . . none of which is technically true. But study their behavior? All they ever do is bask in the sun, waiting to be fed by some lucky chance. Every time you stop by their enclosure in a zoo, you hear some little boy ask, "Are they real or plastic?" What kind of research would it be, sitting for hours in some hot, humid swamp, seeing nothing but wave after wave of mosquitoes and blackflies, waiting for the beasts to move a leg or blink an eye?

Of course, I knew from literature that crocodilians did move sometimes, and that they could do some interesting things: care for their offspring, hunt large mammals, and produce infrasound (acoustic vibrations too low-pitched for humans to hear). But the only things I'd ever seen them do were sliding in the water at my approach and trying in vain to stalk some wading bird. So I went to our library and read the works of Leslie Garrick, a herpetologist who more than forty years earlier had discovered that alligators could communicate by infrasound. He described their behavior during the mating season: the so-called bellowing choruses, head-slapping displays, and other things I'd never heard of.

It was April, the time of year when alligators started mating in Florida and along the coast of the Gulf of Mexico. So I drove to the Everglades, found a roadside pond full of alligators, and waited. The reptiles—thirteen of them, all larger than me—were sleeping on the banks or swimming very slowly through still, tea-colored water. They were black, thickly built, and boring. Nothing happened all day. After sunset, nothing happened either, except now I could see the gators' eyes reflecting my flashlight. I counted those red embers and realized that the lake contained twice as many reptiles as I could see during the day.

The night was full of voices: crickets, tree frogs, toads, nightjars, owls. Everybody was vocalizing except the alligators. The air was hot and humid, as expected. Not as unbearably hot and humid as it gets in the Everglades in the summer months during the rainy season, but still bad enough to make sleeping in a car with windows closed (or your clothes on) impossible. I got in the car, doused myself with insect repellent, opened the windows, and managed to get a few hours of sleep before the repellent evaporated and the mosquitoes rushed in. I woke up, spent half an hour reapplying the repellent and scratching the bites, then got out of the car. And that's when it started.

The lake was barely visible in pink fog. The forest was eerily quiet after the cricket-filled night. The purple sky was crisscrossed with golden jet contrails and lines of high cirrus clouds. The sun was just about to come up. The alligators were all in the water, floating like black rotten logs. Suddenly, the largest one, a beast almost as long as my car, lifted its massive head and heavy, rudder-like tail high above the water surface. He (such huge individuals are usually males) froze in this awkward position for at least a minute, while others around him also raised their heads and tails one by one, until there were twenty odd-looking arched silhouettes floating in the mist.

Then the giant male began vibrating. His back shook so violently that the water covering it seemed to boil in a bizarre, regular pattern, with jets of droplets thrown nearly a foot into the air. He was emitting infrasound. I was standing on the shore at least fifty feet away, but I could feel the waves of infrasound with every bone in my body. A second later, he rolled a bit backwards and bellowed—a deep roar, terrifying and beautiful at the same time. His voice was immensely powerful. It was hard to believe that a living creature could produce what sounded more like a heavy army tank accelerating up a steep rampart. He kept rocking back and forth, emitting a bellow every time his head was at the highest point and a pulse of infrasound every time it was at the lowest. All around the lake, others joined him. They were all smaller, so their voices were higher-pitched and less powerful, but still impressive. Clouds of steam shot up from their nostrils (weren't they supposed to be cold-blooded?). Trees around the lake—huge bald cypresses—were shaking, dropping twigs and dry leaves on churning waters. I stood there, frozen, fascinated, hearing alligators in other lakes, near and far, as they joined this unbelievable show of strength and endurance. For about an hour, waves of bellows and infrasound rolled through forests and swamps all across southern Florida.

Then, gradually, they stopped. It was quiet again. The alligators were floating silently in the black water of the lake as if nothing had happened. I waited for two hours, and not a single one of them moved. Nothing moved there, except the rising sun and flocks of snowy egrets that sailed across the sky on their way from their night roosts to some fish-filled ponds.

I drove home, thinking about what I'd just seen. Both alligators and crocodiles were known to produce sounds (called bellows in alligators and roars in crocodiles) and infrasound during their respective mating seasons. Fossils suggest that these two groups separated about seventy million years ago during the age of dinosaurs, so the spectacle I witnessed was probably even older than that. It was one of the most amazing things I'd ever been privileged to see. It was very easy to observe, yet very few people had ever paid any attention to it. The first description that wasn't total nonsense was written in only 1935, by Edward McIlhenny. McIlhenny was an amateur naturalist, but his book, *The Alligator's Life History*, was way more accurate than those of many professional scientists before him. In the 1960s, Leslie Garrick suggested that these choruses served the same functions as bird songs: attracting mates and staking out territory. But he wasn't sure, and he published only three short papers on the subject. Later, two other zoologists studied it in more detail, but they, as Garrick before them, worked mostly with captive alligators in zoos, not wild ones. It was an area that was all mine to explore. I must be the luckiest zoologist in the history of mankind, I thought.

So the next evening, I was back in the swamps of the Everglades. And the evening after that. And on one of those hot, steamy nights I discovered what no zoologist had ever found before. Alligators did more than just their Jurassic version of bird songs.

They also danced.

2

The Night Dance

Alligator mississippiensis

The hardest bird to hunt is a snipe, for it hides in plain sight.
—Seminole proverb

FLORIDA IS PROBABLY THE BEST PLACE IN THE WORLD to study crocodilians. After being slaughtered almost to extinction, local populations of alligators and crocodiles are now rapidly growing. There are over a million American alligators and a few thousand American crocodiles, plus a few introduced caimans. They share the state with an equally rapidly growing human population, which will probably reach twenty million by the time you read this book. Although it's common to see alligators in urban lakes and irrigation ditches, lethal attacks on humans are surprisingly rare: fewer than twenty people have been killed by alligators in the last fifty years, and nobody has ever been killed by a crocodile in Florida. The statistics have gotten worse in the last decade, in part because the number of very large male alligators began increasing from almost-zero levels since the late 1990s.

Humans don't attack alligators much either: at the time of my research there was no legal hunting, only gathering of eggs for alligator farms and removal of "problem animals" from residential areas. Crocodiles are fully protected by the law. So both species can be very tame and easy to observe.

Encouraged by these statistics, I bought an inflatable kayak and started looking for research sites where I could observe alligators with as little human disturbance as possible. Soon I found two small lakes in different parts of the Everglades, hidden in dense hammocks, which in southern Florida refers to an island of tropical rain forest. Dozens of such islands are scattered across the sawgrass savanna of the Everglades. Some are made up of just a few trees, while others take hours to walk through. Both lakes were filled with alligators. Dry season in the Everglades normally lasts from October to mid-May, so water levels are at their lowest in April and May, just when alligators mate. Any permanent lake usually has lots of alligators at that time. Some gators even dig their own ponds. Only a few feet across when first dug, these "gator holes" are then enlarged and maintained by generations of the reptiles. They are very important for local wildlife, and not just as reliable water sources: the water they hold plus the mounds of excavated soil encourage tree growth, so many hammocks have formed around alligator-made ponds and then expanded.

About fifty miles up the coast from Miami, I found a beautiful place called Loxahatchee National Wildlife Refuge. It had a six-mile-long canoe trail—a narrow, water lily–filled channel winding through sawgrass marshes. There were few alligators there, but I liked having my research locations in diverse habitats.

South of Miami, I chose another site that was probably the easiest place in Florida to study alligator behavior. It's called Anhinga Trail—a half-mile boardwalk near the main entrance to Everglades National Park, at the end of which is a lake. It always has plenty of water, and more wildlife than any place in Florida I know of. During the day it gets lots of tourists. At night there are usually just a few people coming to see alligator eyes reflecting their flashlights.

I made my first discovery on Anhinga Trail during my first week of observations, when I didn't expect to see anything out of the ordinary. Anhingas, weird-looking relatives of cormorants with egret-like necks and heads, were nesting in pond-apple trees in the middle of the lake. Their downy white chicks were already trying their short wings, so numerous alligators floating underneath the trees looked hopeful.

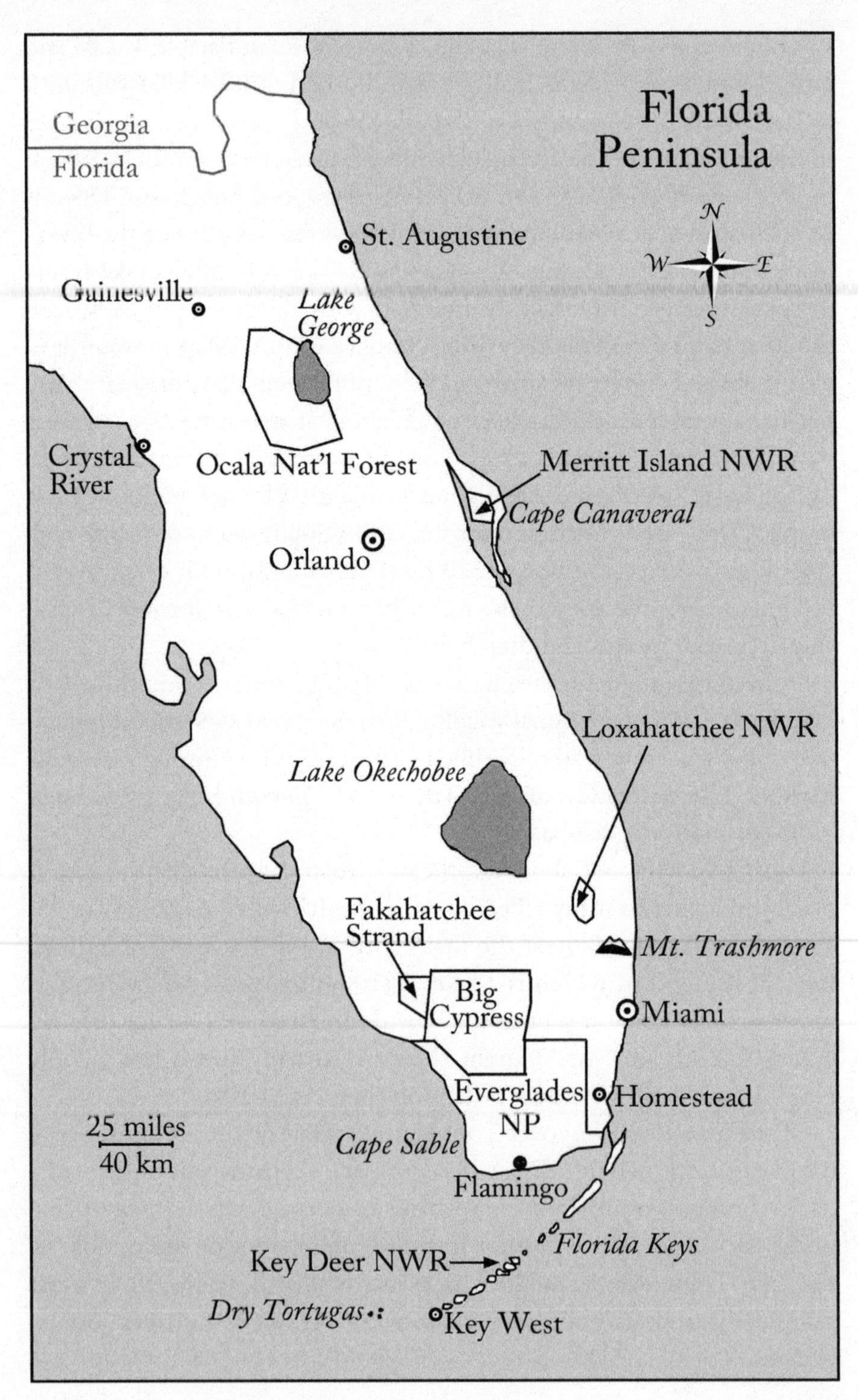
Florida
Peninsula
N
W
E
S
Georgia
Florida
St. Augustine
Gainesville
Lake
George
Crystal
River
Ocala Nat'l Forest
Merritt Island NWR
Cape Canaveral
Orlando
Loxahatchee NWR
Lake Okechobee
Fakahatchee
Strand
Mt. Trashmore
Big
Cypress
Miami
Everglades
NP
Homestead
25 miles
40 km
Cape Sable
Flamingo
Florida Keys
Key Deer NWR
Dry Tortugas
Key West

Two hours after sunset, all the tourists were gone. I switched my tiny headlamp to red light (many animals don't see red color well, so red light is less disturbing for them) and sat on a wooden bench, watching alligator eyes circle below the boardwalk. Soon I noticed that they were all gathering in one part of the lake, and becoming more and more active. Eventually about thirty alligators gathered in an area less than sixty feet across, swimming like crazy, splashing, hissing, slapping their heads and tails, occasionally getting into brief but violent fights. Some of them would form pairs, then break up again. New ones were arriving every few minutes, alone or already in pairs, smaller females following their males. Others were leaving, but many remained in that small area until dawn. Then all swimming and fighting and splashing ceased, and the lake was quiet again. After the sun came up, the remaining alligators bellowed in chorus, and crawled onshore to bask for the rest of the day.

I saw gatherings like that almost every night for a few weeks. In lakes, alligators would choose one part of the lake; in Loxahatchee, they did it in a large canal, always at a different location, but within the same area.

What were they doing? To me it looked like village dancing parties, where people would come, alone or with their spouses, to socialize, have fun, and, in the case of singles, search for mates. I looked through literature, but such "dances" weren't even mentioned anywhere. Local naturalists I consulted had no idea what I was talking about.

This was astonishing. The American alligator is probably the most-studied reptile in the world. Thorough accounts of its natural history have been written by famous naturalists since the eighteenth century. At least a thousand papers dealing with its anatomy, physiology, population demographics, and, of course, behavior have been published. It's unbelievably easy to observe: any Miami resident can get to Anhinga Trail in less than two hours. But nobody had ever noticed that alligators "dance" at night. How was that possible?

Eventually, I understood why I got so lucky. These "dances" are difficult to observe in captivity. Captive alligators are kept together all year and know each other well, so there's no need for displays and fights. And even if you did see them swimming and thrashing around at night, you'd have no idea what was going on. I'm sure tourists, fishermen, and hunters have witnessed "dances" on occasion, but they probably didn't pay attention, or didn't realize it was something unusual. Although everybody knows alligators are primarily nocturnal, the few people who studied them in the wild were mostly interested in nesting and mother-offspring interactions, and observed them only during the day.

Humans are a diurnal species. Even some experienced naturalists feel uncomfortable being in the forest or in a swamp at night. I know a few field biologists who have lived in the jungle for years who never set foot outside their cabins after nightfall.

I was lucky to be born without this subconscious fear of the dark. For me, night is the most fascinating time. I can't sleep well on the nights of full moon; I can walk through dark forests and deserts or swim over sleeping coral reefs forever without getting bored, and I've learned that there's usually much more interesting stuff to see outdoors when it's dark.

I grew up in the central part of Moscow, an imperial capital of ten million people. By the age of four I was interested in wildlife, but there was almost none around. In the summer we'd go to the country, but for the rest of the year there were only pigeons, sparrows, and, after the snow melted, a few insects inhabiting lawns and small parks. My mother was a normal person, not a naturalist. Although she was very sympathetic to my obsession, she could offer little help. She was more concerned with my general education, so when I turned five she took me to the Bolshoi Theatre to see Tchaikovsky's *Swan Lake*.

It's a long ballet and was scheduled to end at midnight—much later than I was normally allowed to stay up. Throughout the performance, I kept telling her, "I wish it lasted longer. I wish it wouldn't end for another few hours." Of course, my mother was happy to hear this. "Isn't it great that he already has such appreciation of classical music!" she thought. But, as I kept repeating that I wished the ballet to last for as long as possible, she finally asked me:

"Why do you want it to last for so long?"

"Because if it ends late enough, maybe we'll see some bats on the way back."

By the time I was seven, she had gotten so used to accompanying me on my nighttime walks through suburban forests that she began to like them. It was still Soviet times, and the Russian countryside was relatively safe. I never managed to infect her with my passion for wildlife, but she learned to enjoy travel to remote places, and nature in general.

Being outdoors at night so often has allowed me to see some animals that few, if any, biologists have ever seen, such as the Andean cat and the Congo bay owl. My luck with alligator "dances" was probably due to my nocturnal habits as well.

I learned by observing a large female with a brood of twelve babies that lived in a pond in Loxahatchee that the alligators apparently considered these gatherings very important. Alligator hatchlings, although smart and agile, are extremely vulnerable; they need their mother's protection to survive in a world full of predators. But this female had other priorities. Every night she'd leave her babies and swim to the large canal two miles away where "dances" were taking place. She wouldn't return until late morning. In three weeks, half of the tiny baby alligators disappeared. I saw one being taken by a great blue heron, but raccoons and bald eagles could also have caught a few.

Figuring out what exactly was going on during those "dancing parties" was tricky. Recognizing individual animals was all but impossible in the melee. I couldn't even tell males from females unless I observed the animal in question mate or bellow (only males accompany bellows with infrasound), or it was over eight feet long (females don't usually grow that big).

Another problem was that virtually nothing was known about the private lives of wild alligators. It was believed that they were promiscuous, and that each would mate with multiple partners if given a chance. A few years after my discovery, a study was published showing that, despite being polygamous, alligators of both sexes have preferred partners with which they mate year after year. But this only further complicated things.

It wasn't even clear if males were territorial: often they'd tolerate others dancing, bellowing, and courting females a few feet away. Nevertheless, almost every night I witnessed gruesome fights that occurred for no obvious reason. One young male lost his upper jaw in a fight, and died a week later. A few were missing their front feet, but I never saw an alligator missing a rear foot. I think it's possible for an alligator to survive without a front foot, but not without a rear one, because rear feet are used for steering while swimming (the only source of propulsion is the tail, in which more than half of an alligator's formidable muscle power is concentrated).

Courtship and mating could happen during the "dances" or at any other time of night, and sometimes in the morning. When I finally started to recognize a few individual animals at my study sites, I learned that both males and females could initiate the courting. One, two, or three males would follow a female for a while; on most occasions she'd rebuke them by hissing, growling, or slapping her jaws. But sometimes one suitor would persist, swimming beside her or following her as she swam in tight circles, touching her with his nose or chin. Alligators have musk glands on the underside of their lower jaws, and some have suggested that touching makes it easier for the partner to smell the musk. But I suspect that there is something else to this chin-touching—that probably the chin is particularly sensitive sexually. A recent study has found that an alligator's massive jaws are more sensitive to touch than human fingertips, which explains how mother alligators can carry their hatchlings without harming them.

If females initiated the courtship, they were much more straightforward. When a large, strong male bellowed, his display sometimes had a spectacular effect on nearby females: they'd rush to the male and place their chins on his back. This usually led to sex after just a few minutes of nose-touching, although sometimes a macho male would totally ignore the ladies' advances.

One morning I was watching alligators in two small ponds. In the pond on the right side of the trail were eight large gators; the pond on the left side contained six smaller ones, four to five feet long, the size at which wild alligators start mating. An hour after sunrise, the eight large alligators in the pond on the right side bellowed in chorus. As soon as they finished, the "teenagers" in the other pond did their own version of a bellowing chorus, weak and somewhat pathetic, but very enthusiastic. Then a small female cautiously approached a young male and touched him with her chin. They spent at least an hour caressing each other with their chins and feet; I've never seen foreplay so tender and patient. Then they made love. It was clear from their clumsiness that for both it was the first time.

And it was also perfectly clear that, no matter how deeply touched I was by their romance, I had to make sure none of my guesses would ever show up in a scientific paper.

A scientist trying to study the behavior of wild animals is not supposed to get emotional. Once you get too personal, you invariably start assigning human feelings to your study subjects. This subconscious humanization can lead to grave errors in interpreting what you see. You shouldn't forget for an instant that every species is different, even primates, to say nothing of creatures as distinct from us as crocodilians. But staying completely objective and impartial can be very difficult, almost impossible.

I was totally unprepared for the level of behavioral complexity I was observing. It's known that alligators and other crocodilians belong to a highly evolved group of animals; their closest living relatives aren't other reptiles, but birds. Some fossil and physiological evidence, such as having a four-chambered heart, seems to suggest that crocodilians have descended from bipedal, warm-blooded, terrestrial predators. People who have raised crocodiles from eggs have succeeded in training them to do very complex tricks, and some of those human-raised crocodiles have lived for decades in residential houses side by side with the owners' children and grandchildren, never doing the little kids any harm. Still, who'd ever think of crocodiles or alligators as being smart? Aren't they just mindless killing machines?

3

Learning the Basics

Alligator mississippiensis

If we knew what it was we were doing, it would not be called research, would it?
—Albert Einstein

A MONTH HAD PASSED. I was getting exhausted from long nights and mornings of observation (the alligators invariably "danced" at night and "sang"—bellowed—in the morning), endless drives between my study sites, and going through piles of old scientific journals in futile attempts to understand what was going on. I also had to teach undergrads and take classes two days a week.

I realized that studying the "dances" thoroughly would require much more time and manpower than I could provide. It would have to be a long-term study involving paint-marking most alligators in some area. I was expected to graduate within five to six years; the alligators' mating season was just six weeks long. My only hope of getting financed was to apply for research grants. Grants available to graduate students were difficult to obtain and mostly very small.

But for now, I was having a great time. After so many years, I could finally do some independent research, wandering into a completely new area of natural history, uncovering secrets that had been hidden since the time of dinosaurs. Most alligators spent the hottest hours of the day basking semicomatose onshore, giving me a break I could use to get some sleep or to dart back to town for a shower and some ice cream without too much risk of missing something interesting.

And, of course, spending so much time in the forests and swamps of southern Florida meant seeing so many other wonderful things.

One of "my" lakes was inhabited by a family of otters. At first there were four, but as the water levels kept falling, two more walked in along a dried-up stream. The new arrivals were greeted in a very friendly way, probably because the otters were all relatives or old friends. No, I'm not falling into the trap of humanizing here—such long-term friendships are well documented for many species of carnivorous mammals. Watching the otters was nonstop fun. In just three days they grew so tame that sometimes they'd play with my toes or jump into my kayak and take rides. They caught more fish than they could eat, and accidentally provided me with a few nice meals of grilled catfish. But they particularly enjoyed playing with small alligators. Being much faster, otters could tease them by almost touching noses with them, nipping on their tail tips, or splashing water in their faces. Most alligators would submerge to avoid harassment, but one five-foot youngster couldn't stand it and lunged at the otters every time they got close. So the otters focused their attention on this one most of the time. They easily avoided its jaws, until one of them slipped on a steep bank and was instantly grabbed by the gator.

I expected to see a terrible death of a beautiful animal that by now I considered a personal friend. The alligator stepped back, firmly holding the wriggling otter in its teeth, and lowered its head, pulling the otter underwater (drowning is the method crocodilians commonly use to kill mammalian prey). Then, suddenly, it raised its head again and opened its mouth, releasing the otter completely unharmed. Why wasn't the otter killed? Did the alligator get a painful bite? Or was it also playing? I will never know. It was one of many observations suggesting that crocodilians might be much more intelligent than anyone could ever imagine that I was to collect over the years, though most were too isolated to publish in a scientific journal.

By mid-May, the rainy season was approaching. Small canals in Loxahatchee were so overgrown with water lilies that it was difficult to push my kayak through. White swamp lilies opened their ghostly stars between the black roots of bald cypresses. Towering cumulus clouds formed every afternoon above the parched savannas, but not a drop of rain had yet fallen in the Everglades. In the northern part of the swamps, where fertilizer-rich runoff from sugarcane fields had caused sawgrass to be replaced with impenetrable foxtail thickets, huge wildfires were spreading, often blanketing Miami with smoke. Sunsets were unnaturally colorful, but sunrises were clean, tender, and calm, with mists the color of pink seashells.

This was one of the driest springs on record in southern Florida. "Otter Lake" soon dried out completely. My otters ate all the remaining crayfish and frogs, then broke apart a huge stump in the center of the lake to get to the few water snakes hiding inside, then left. Alligators, too, had to crawl away, leaving deep ruts in the mud. Elsewhere, alligator populations were overcrowded and stressed out. Three people were killed by alligators in Florida within just two weeks. Two cases particularly horrified the Floridians because they happened relatively far from water: a girl was killed while jogging along a dike, and an old woman while watering her garden. At the time, it was widely believed that alligators never hunted on land; it would take me two more years to prove this belief wrong.

Finally, the rains arrived. Brief, spectacular thunderstorms were crisscrossing the open plains, dousing fires, reviving plants, saving thirsty animals. The humidity shot up, and the afternoon heat became unbearable. Mosquitoes, blackflies, no-see-ums, yellow flies, gnats, and horseflies ate me alive. It got cooler only during thunderstorms—sometimes so cold that to stay warm I had to float neck-deep in the water under my overturned kayak.

The inflatable kayak had proved to be one of the best things I'd ever bought. I could carry it in a backpack, penetrate even the shallowest channels with it, sleep in it comfortably, and glide quietly through the swamps, approaching courting alligators to within an arm's length.

It would take some time for the rainwater to creep south from the lakes of central Florida and fill up the Everglades. Despite the rains, some ponds were still drying out. A few times I saw alligators engaged in so-called cooperative feeding: a few dozen would gather in a shallow lake filled with fish and spend the night swallowing one catfish after another, until almost nothing was left.

They stopped dancing and almost stopped bellowing. I saw many of them in pairs, and still observed courtship occasionally, but generally there wasn't much to see. It was time to pause and decide what to do next.

I had more than ten months until the next alligator mating season. The semester was over, so I didn't have to teach or attend classes. There was nothing to do in the Everglades either, though I still checked on my study sites once a week, just to see what was going on.

By late May, males were mostly relaxing after two months of crazy nightlife, while some females were already busy building their nests, large piles of dry leaves and branches. Alligators take care of their eggs in the same way primitive turkey-like Australasian birds called megapodes do, and most dinosaurs probably did it that way as well. They put their eggs inside the nest mounds and let the sun and rotting vegetation provide the heat for incubation. Sometimes the female would seem to shade the nest with her massive body or to wet it with water dripping from her belly, but attempts to prove that females can intentionally regulate the temperature inside the nest have failed so far.

Every nest has a slightly different inside temperature, a variation that is important in a surprising way. The sex of hatchlings in alligators is determined by temperature: the hotter the nest, the more males are born. In crocodiles it's even more complicated, as extremely high temperatures also produce females. But alligators are subtropical animals (crocodiles are mostly tropical), and their eggs can't survive extreme heat that well. Even adult gators get stressed by the summer heat in the Everglades, which is near the southern limit of their range.

It is anybody's guess why crocodilians evolved such a strange method of sex determination and how it is going to be affected by global warming.

All this was very interesting, but I didn't want to look into nesting, hatching, and parental care. There were already enough people studying these aspects of alligator biology. Still, I knew I had to broaden the scope of my research if I wanted to get my PhD in six years rather than six decades. If I studied just the "songs" and "dances" of American alligators, I'd be able to collect data for only two months a year.

That's how my six-year journey around the world began. It seemed simple at first. I just decided to look at other species, and to plan some kind of a comparative study of "singing" behavior. And I still had a year to submit my dissertation proposal.

At that time there were thought to be twenty-three species of crocodilians. The number is slowly growing as new data from DNA studies trickle in and some species are getting split; by the time you read this book the widely agreed-upon figure will most likely be approaching thirty.

I soon found that quite a lot was known about the genetic makeup, anatomy, and nesting of crocodilians but very little about their mating habits. When I was a schoolboy in the Soviet Union, the chapter on reproduction in my *Human Anatomy and Physiology* textbook began with "After a sperm cell reaches an egg cell . . . ," modestly omitting everything that led to fertilization. Much of the literature on crocodilians had a similar blank spot. I found published information on mating-season sound-producing behavior for only six species, and most of this information came from captive animals and was very fragmentary. It seemed that people studying crocodilians had almost completely ignored that aspect of their natural history, being more interested in what happened a month or two after mating: nest-building, hatching, parental care—things that are much easier to observe in captivity.

Although crocodilians are most closely related to birds, they have traditionally been studied by herpetologists (people studying reptiles). Snakes, lizards, and turtles are usually too secretive to allow any regular observations, and most such studies are done on birds and large mammals. So herpetologists aren't used to studying animal behavior in the wild. And they are even less used to studying social behavior, because an old (and completely wrong) dogma states that reptiles are mostly solitary creatures.

Since published data on crocodilian "songs" was so limited, I tried to learn more about the subject by contacting researchers working with crocodiles. Mostly, I drew a blank. Even zoo curators were often unaware of their animals' behavior because courtship-related activities took place outside normal working hours. And some scientists and reptile breeders simply didn't want to share information.

But a few people were very helpful. Liz von Muggenthaler, a pioneer researcher of infrasound communication by giraffes and cassowaries, told me about one old model of Sony tape recorder that, probably by chance, was capable of recording infrasound. I found a used one online for just a fraction of what infrasound-recording equipment normally costs. Tapes for that model were no longer available, but I managed to get a couple from Sony headquarters in Japan.

Steve Irwin, the Australian conservationist made famous by his television show *Crocodile Hunter*, enthusiastically invited me to visit him in Australia to study saltwater and Johnston's crocodiles. Unfortunately, I never got to meet him in person: he was tragically killed by a stingray the same year.

Another scientist who replied to my email was John Thorbjarnarson, a herpetologist at the University of Florida. He was in his late forties, but he had already become something of a patron saint of endangered crocodiles and alligators. In the twentieth century, many crocodilians were pushed to the brink of extinction by hunting and habitat loss. Nowadays all species are bred in captivity, and most are common in at least parts of their former range, but some are still far from being safe in the wild. John was working with Orinoco and Cuban crocodiles and the Chinese alligator, some of the rarest. Without his incredible organizational skills and perseverance, these species probably wouldn't be around by now.

John invited me to visit him in Gainesville, a few hours' drive north from Miami. Modest, friendly, and easygoing, he was a walking encyclopedia of all things crocodilian. In just two hours of discussing my research plans he made so many interesting suggestions that following up on them all would have taken a lifetime. For now, I had to choose one, and I decided to look at the closest relatives of American alligators.

4
Dragon Reborn
Alligator sinensis

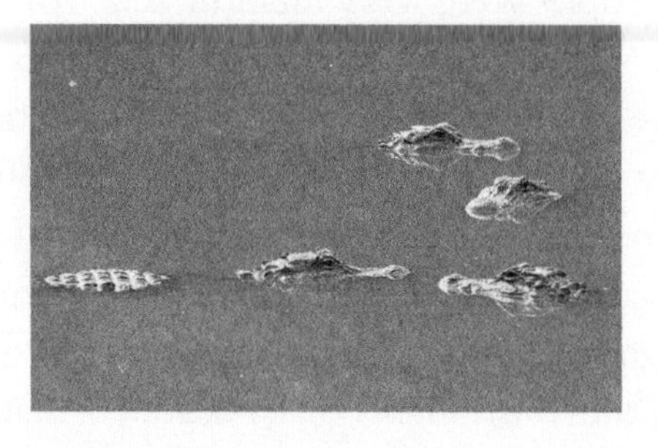

A good traveler has no fixed plans, and is not intent on arriving.
—Lao Tzu

THE ONLY OTHER ALLIGATOR SPECIES IN THE WORLD is the Chinese alligator. These small, harmless reptiles once inhabited much of East China, but by the turn of the twentieth century there were just a handful of them left in one tiny part of the immense floodplain of the Yangtze River. They survived in a few village ponds separated by miles of rice paddies and cornfields, so they weren't breeding anymore.

There were still hundreds in captivity in a special breeding center in China; the problem was finding places to release them back into the wild. Most of East China is now one endless human settlement, a mix of rural and urban areas, but continuous. Only sacred mountains, densely forested and sheltering plenty of wildlife, rise above this sea of humanity. But alligators need wet lowlands, not mountains.

John was trying to solve this problem, and was spending a lot of time in China. He gave me email addresses of some Chinese alligator researchers, who invited me to visit the breeding center. Due to freezing winters and cool springs, Chinese alligators don't start mating until mid-June, so I had some spare time to travel around China.

I had been to China once before, in 1993. Russia was falling apart then, and obtaining foreign currency was almost impossible, so I had to travel for four months on a budget of $200, sleeping in irrigation ditches and feeding on whatever people would share with a wandering sage. I had a self-made letter of introduction (translated for me into Chinese by a friend's friend) claiming that I was a great Russian writer and the best friend of the Chinese people. This piece of paper, covered with fake official stamps, saved my life on a few occasions. I kept traveling until I almost starved to death, seeing unbelievable things that have since disappeared forever, like a mountain valley where no other human had set foot since at least the last Ice Age. Just a few years later, a highway was blasted through that valley. I was very young and . . . let's say, motivated.

Now that I was a respectable researcher, I didn't have to hitchhike all the time and could pay for my meals, but I was still going to have some adventure. That semester I was dating a Chinese girl whom I'll call "Li" here, although her real name sounded even more musical. Exquisitely beautiful and delicate, she looked like a porcelain figurine from the Tang dynasty, but in reality she was a talented chemistry student and loved spicy jokes. She had just been offered a job in China and was about to return there. Her father, a high-ranking Party official, was very unhappy about her dating a foreigner, especially an ex-Russian.

The USSR/Russia has betrayed Chinese Communists three times. First in 1945, after the Red Army wrestled Northeast China from the occupying Japanese and Stalin turned that territory over to Mao's archenemy, Chiang Kai-shek. Stalin's motives for mistrusting fellow Communists are still unclear. The second time in the late 1950s, when Khrushchev abandoned Stalin's murderous policies and expected Mao to do the same. When that didn't happen, Khrushchev canceled the colossal aid program, leaving China full of tractors with no spare parts, factories with no equipment, and half-built bridges. The third time in 1991, when the Soviet Union officially abandoned Communism (the fact that China was already a capitalist economy

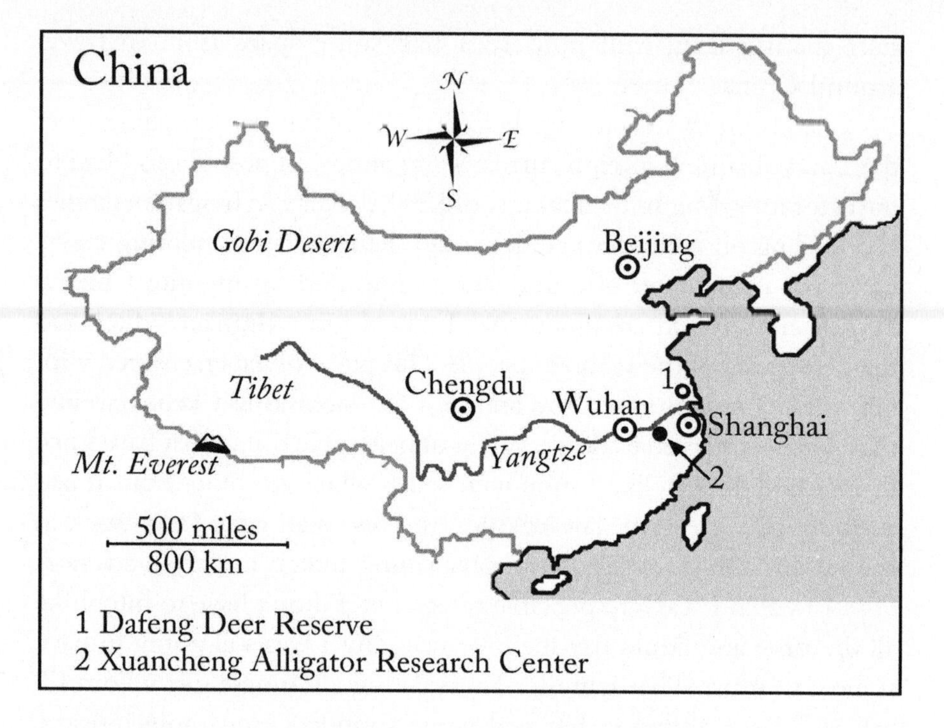

didn't matter). The ordinary Chinese don't take these grievances seriously, but the old guard of the Party did at the time. No wonder Li's father refused to meet with me. However, he agreed to let us use his huge SUV for part of our trip.

So the plan was for me to fly to Shanghai, make my way to central China, meet Li, drive to Tibet and the Gobi Desert, then get her home and go back east to do my alligator studies.

I was prepared to see a country changed since my visit in 1993, but what I saw was still unbelievable. When my plane landed and a high-speed train carried me to Shanghai, it was in the middle of a thunderstorm. Futuristic black shapes of well-spaced skyscrapers were piercing the flying clouds, and lightning bolts were dancing between them. It all looked like the awakening scene from *The Matrix*.

China was almost impossible to recognize. In six weeks nobody ever tried to pick my pockets. In 1993 it happened almost daily, and the country generally felt very unsafe: truck drivers carried guns, bazaar crowds would regularly run around chasing thieves, public executions of criminals were conducted in city squares. Now you could sleep on a train without tying your backpack to yourself or putting your shoes under your head.

One major change was the rise of in-country tourism. Almost everywhere we went, there were hundreds of tourist buses flocking to all possible attractions. Small tribal villages were being replaced with huge fake ethnic towns built of concrete. Every provincial capital now had an excellent museum; the ones in Chengdu, Wuhan, and Shanghai were among the most interesting I've ever seen. Buddhism and Islam, once oppressed, were now promoted as tourist draws, and Lamaist monasteries were all sporting recently painted walls and gilded roofs.

Roads were either amazingly good or being rapidly built. Even in the most remote mountains we encountered worker crews of hundreds, constructing modern highways, bridges, and tunnels. On the plains, most roads had huge tolls, but we didn't pay them: just before each tollgate Li would get behind the wheel and pretend to be the driver-interpreter for an immensely important American visitor (me).

One big change I was particularly glad to see was in people's attitude toward the environment. Drivers now tried to avoid hitting feral dogs rather than chase them. In street markets there was almost no wildlife for sale, except for poached saiga antelope horns from Kazakhstan and obviously fake tiger pelts. We saw virtually no logging and a lot of tree planting. North from Shanghai, I visited a tiny nature reserve established for the reintroduction of Père David's deer, extinct in the wild. To create this little patch of green, the government had to resettle tens of thousands of people.

Unfortunately, it was often impossible to save the last wild places from development. Much of the Tibetan Plateau was terribly overgrazed, and desertification in Inner Mongolia was so bad that neighboring provinces looked like the Dust Bowl. Native species were going extinct at a horrifying rate. An expensive breeding center for the endangered Yangtze River dolphin had been built too late, as no surviving individuals could be found. At the time of my visit, the center was housing small, unbelievably charming Yangtze porpoises, of which a few hundred still remained (porpoises are small relatives of dolphins, differing mostly in teeth shape).

But some things didn't change. Cheaper railroad cars were sometimes packed so tightly that people had to ride standing for thirty or forty hours, wearing diapers since the bathrooms were also packed. All signs in English still had errors. Hair on my arms still made me very popular, especially among young women (for particularly good-looking ones I pulled down the collar of my T-shirt, showing the hair on my chest—that made them swoon). Sometimes Li pretended that she didn't understand the local dialect, and then translated for me the discussions the women had about us (they mostly tried to guess the dimensions of my unseen body parts and wondered how it felt to be close with someone so hairy).

Li also taught me Chinese card games and gave little history lessons. Chinese history is as rich in interesting story lines and anecdotes as the history of Europe, if not more. It was a really sad moment when we had to part ways . . . but at least I managed to return her father's ultra-expensive SUV without a scratch despite being through a sandstorm, a snowstorm, and a few difficult mountain passes.

The alligator breeding center was located on the outskirts of Xuancheng, a small provincial town. Large ponds were filled with thousands of Chinese alligators—small, stocky, rather cute animals that looked a bit like the inflatable toy crocodile I used to have as a kid. A part of the center was set aside as a nature reserve of sorts. Lakes there were used as breeding pools, and a few females were already guarding their nests. The largest lake looked empty, but it had a few alligators that weren't fed and behaved like wild animals.

At that particular moment the population of wild Chinese alligators was at its lowest point in the history of this ancient species, which had evolved millions of years ago. There were about twenty animals left, all stuck in tiny village ponds and no longer breeding. I managed to find one such animal in a valley not far from the center, but it was obviously stressed and didn't "sing."

Starting in 2003, an ambitious reintroduction project was launched in East China, and by 2009 the reintroduced alligators had started breeding in the wild. The project took a tremendous effort from John Thorbjarnarson and many of his colleagues in the US and China, as well as some large-scale official support. Whatever your opinion of the current Chinese government is, one thing you have to admit: they take rare species conservation seriously.

I was reluctant to study alligators in captivity, because I already knew that captive crocodilians, even if kept in large enclosures, behave differently than truly wild ones. But I had no choice. So I started watching the alligators in the largest lake. They were so shy that I had trouble finding them; they were apparently spending a lot of time in the well-hidden burrows that this species uses to survive long, freezing winters and to escape hunters.

It was obvious that I got there just in time. Those I did see were constantly courting and bellowing. I tried to learn whether they produced infrasound vibrations, but didn't notice any sign of it. (Later, Chinese researchers found that the largest males, over five or six feet long, can do it, but their infrasound pulses are very short. It was an important discovery–I'll return to this later.)

As I talked with the center personnel, I made another discovery. There was a Chinese student interested in alligator behavior! And he was studying exactly the same things I was beginning to focus on! Normally it's an unpleasant surprise for a scientist to find out that someone else is working on the same subject. But I was relieved. There was so little known about crocodilian signals, and the field was so extensive and untouched that I was happy to share it with someone. Now I could let Xinhuan Wang, my new friend, gather information on the Chinese alligator, while I'd be busy with the remaining twenty-odd species.

Xinhuan showed me around. He was still in his teens and looked like a schoolkid, but he proved to be a capable researcher. Later he coauthored with John Thorbjarnarson a beautiful book on all aspects of Chinese alligator biology, in which he described everything I wanted to know about the social behavior of this "toy gator." I was grateful for having so much data to use. But the book would come out only three years later. For now I was stuck with old papers by Leslie Garrick.

In their classical works on crocodilian bellows and roars, Garrick and his coauthors described two kinds of displays produced by American alligators in spring. One is the bellowing display, the impressive sequence of bellows and infrasound pulses that I enjoyed observing in the Everglades so much. It's usually performed in choruses, and joined by females (they also bellow, but without infrasound). The second is the headslapping display: it looks similar, but instead of bellowing the animal would slap the water surface with its heavy head, producing a sharp splash like the ones beavers make with their tails when alarmed, just more powerful. Males accompany it with an infrasound pulse. Headslaps are made only once or twice in a row, and never in choruses.

Garrick and his colleagues described both kinds of displays as being used regularly by alligators in central and northern Florida. But in the Everglades, I almost never observed headslaps. Now I noticed that Chinese alligators hardly ever headslapped. Why such a difference? And why do alligators have two similar displays rather than one?

There was another mystery. Alligators bellow a lot, but crocodiles seldom roar, although they are also known to headslap and produce infrasound. Why? Garrick thought that this difference was due to habitat preferences. Supposedly, alligators live in marshes where they can't see each other, so they need to communicate by sound. I didn't buy this. The American alligator and most crocodiles are *habitat generalists*: they'd settle in almost any body of water available, from overgrown marshes and tiny ponds to large lakes and rivers. There had to be some other explanation.

I kept thinking about these questions, but it was difficult to come up with any possible answers because so little was known. How do alligators behave in other parts of their range? Do all crocodiles roar rarely, or only the two species that Garrick studied? What displays do other crocodilians—caimans and gharials—have?

I went back to Shanghai and spent a couple days exploring this pleasant city, which makes New York look like an ugly, overbuilt provincial town. Unfortunately, I had to waste some of this time in a dental clinic. A local candy I tried was so sticky that it pulled a filling out of my tooth. Filling replaced, I flew to Miami, where I had a lot of interesting problems to tackle and two months of summer to spend on them, but no research plan and no ideas on what to do next. There were plenty of good questions; the difficult part was coming up with practical ways to find the answers.

An ancient Chinese "magic bowl," Pingyao, China. Rubbing the edges generates infrasound and Faraday waves.

5
Missing a Shot
Caiman yacare

It is the ancient wisdom of birds that battles are best fought with song.
—St. Francis of Assisi

THE GREAT NATURALISTS OF THE PAST DIDN'T HAVE TO WRITE thesis proposals. Charles Darwin spent five years circumnavigating the globe on HMS *Beagle*, observing, collecting data, learning. Then he took twenty years thinking about his observations, and eventually wrote *On the Origin of Species*.

Nowadays the process is supposed to work backwards. You come up with a hypothesis that can be tested quickly and inexpensively, and is simple and conventional enough to sell to various committees and boards. You design an experiment to test it. Then you write a proposal or a grant application, which is supposed to contain a detailed research plan. If it gets accepted, you are finally free to go into the field and have a look at your study subject. You have to pray that everything goes according to your plan and you get a positive result. Everybody knows that negative results are equally important for science, but negative results are virtually impossible to publish or defend as a thesis. If you are lucky and you get what you want, you then spend years writing it up, sending it out to journals (each with its own complicated requirements), arguing over it with anonymous reviewers who often don't even bother to read your text well and keep coming up with irrelevant complaints, and waiting for it to be published. Your actual research and thinking is sometimes less than a quarter of the total time spent. This system not only stimulates scientists to conduct biased research (intentionally or not), it often ensures that people generate hypotheses before they get to know their subject well. No wonder we don't get works like *On the Origin of Species* anymore.

To make the process less illogical, I could do a so-called "pilot study" to become more familiar with whatever I was going to work on before submitting the thesis proposal. There was no funding; I was supposed to teach eight months a year to get my stipend and received no money whatsoever in the summer. Miami is not a cheap place to live, so with just the stipend it was all but impossible to pay for travel outside Florida. Few graduate students conduct truly original fieldwork abroad nowadays, preferring to study something nearby or to slave on slices of their advisors' research. I didn't have the latter option. My advisor, Steve, was always ready to share wisdom or help work out details, but he wanted me to do all the major steps independently. I have to admit I liked it that way.

Fortunately, I had some outside financing, so to speak. In the early 1990s I happened to be one of the first Soviets to start traveling by myself far from home, going to China, Latin America, and other places that previously seemed almost as inaccessible as other planets. When I came back, I typed my travel diaries on one of the first personal computers in Moscow. These travelogues became very popular, and within a few years, hitchhiking to other parts of the world developed into a subculture involving thousands of people in the former USSR, where almost nobody could afford conventional travel. It wasn't an easy lifestyle, and some of my friends perished in Afghanistan or Congo or went through rather unpleasant adventures. Imagine hitchhiking from Moscow to Namibia, planning to return by a Russian cargo ship from Cape Town, and having instead to hitchhike all the way back because South Africa won't give you an entry visa (it took immigration officials worldwide many years to become less paranoid about Communist infiltration). My travelogues and a few wildlife guides I coauthored are still being republished, so a few hundred dollars per year keep trickling in.

I went to London to see my mother. She lives in Moscow, but we prefer to meet elsewhere because Russia is not an inviting place. She brought me some money from Russian publishers, and we immediately blew half of it by renting a car and driving around England and Scotland. I owed it to my mom. She had to go through a lot while I was traveling and writing those travelogues. There was no Internet at the time, Russian telegraph was very unreliable, and sometimes she had to worry for weeks, not knowing whether I was alive or dead.

Now I could afford a short field trip. I tried to find out which crocodilian species had its mating season in August, but various books contradicted each other. (I later realized that authors often confused mating and nesting seasons, which are a month or two apart.) Finally, I decided to try Bolivia, which wasn't too expensive to fly to from Miami and had five or six caiman species. There was a good chance at least one of them would be mating. Caimans are related to alligators, but, unlike alligators and crocodiles, they live only in the New World, mostly in the tropics of South America.

You'd think Bolivia is an exotic place, but it doesn't seem so when you land. I arrived from colorful, beautiful, charming Britain, its little hedged fields looking like a stained glass window from above. I had a short stopover in Miami with its gorgeous tropical clouds and turquoise seas. And then I landed on the Altiplano—the intermontane plateau of the central Andes, very high, dry, and cold, a brown expanse of dead grass and brick villages.

I squeezed into a bus going to La Paz, the country's capital. We spent a few minutes making our way through rather unpleasant suburbs, and suddenly the Altiplano ended, and all first-time visitors gasped.

The land was falling in front of us into an immense valley more than a mile deep. Imagine the Grand Canyon densely built over: concrete high-rises and cathedrals along the bottom, brick honeycombs of working-class *barrios* climbing up the walls all the way to the rim and spilling over it. This chasm was getting deeper and deeper as it

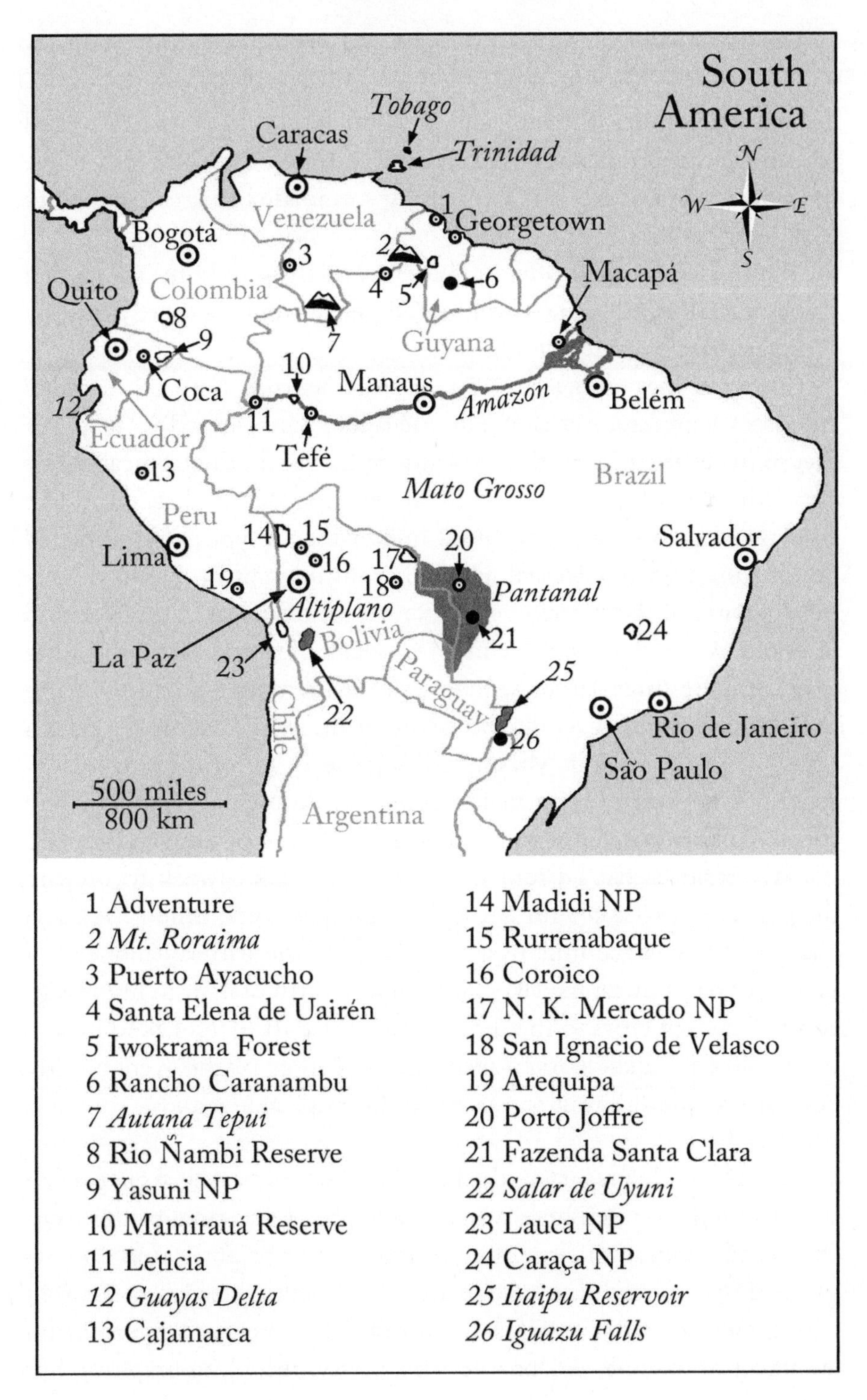
South America
N
W
E
S
Tobago
Caracas
Trinidad
Venezuela
Georgetown
Bogotá
Colombia
Quito
Macapá
Guyana
Coca
Manaus
Amazon
Belém
Ecuador
Tefé
Mato Grosso
Brazil
Peru
Lima
Salvador
Pantanal
Altiplano
Bolivia
La Paz
Paraguay
Chile
Rio de Janeiro
São Paulo
500 miles
800 km
Argentina
1 Adventure
2 Mt. Roraima
3 Puerto Ayacucho
4 Santa Elena de Uairén
5 Iwokrama Forest
6 Rancho Caranambu
7 Autana Tepui
8 Rio Ñambi Reserve
9 Yasuni NP
10 Mamirauá Reserve
11 Leticia
12 Guayas Delta
13 Cajamarca
14 Madidi NP
15 Rurrenabaque
16 Coroico
17 N. K. Mercado NP
18 San Ignacio de Velasco
19 Arequipa
20 Porto Joffre
21 Fazenda Santa Clara
22 Salar de Uyuni
23 Lauca NP
24 Caraça NP
25 Itaipu Reservoir
26 Iguazu Falls

snaked away from us, descending into dusty Andean foothills. But we couldn't see the Amazon lowlands. Instead, the backdrop to the unbelievable canyon was the icy ridge of Illimani, soaring eight thousand feet above the rim of the Altiplano, and fourteen thousand feet above the lowest parts of the city. The setting sun made the great mountain look like a wall of purple flame.

This place is called *Mirador el Alto* ("Heights Lookout"). I spent a week in La Paz, and every evening I'd take a minibus there from the central square to watch the sunset.

Other things didn't go well. Before leaving home, I emailed Alfonso Querejazu, my Bolivian colleague, who offered to help. We met in his little office in the far south of La Paz, in the tropical desert zone (there are four climate belts in the city). He looked more like a wiry, tanned, mustachioed forest hunter than a typical bespectacled South American intellectual. He had spent most of his life in the remotest parts of the Amazon and knew a lot of things you never find in books. As we discussed my plans in detail, I learned that the mating season of the three larger species of caimans wasn't going to begin until three months later, at the onset of the rainy season, by which time I had to be back in Miami teaching classes. As for the two smaller species, they were so rare that nothing was known about them. My trip to Bolivia was about to become a total waste of time and money.

We decided that I'd return next year. It took a week to prepare the paperwork to apply for a research permit. All that time I lived in a cheap hostel, entertaining myself with trips to the arid badlands east of La Paz, or to frigid alpine lakes and tundras north of it, or to impressive ruins of ancient cities to the west, around the dark, deep Lake Titicaca, or to colorful Indian markets near the city center. Nights were bitterly cold, and I couldn't wait to escape to the tropical lowlands.

One day I couldn't return to the hostel because the streets were packed with excited crowds. There was a parade of folk dancing groups from all over the country. Teams of young dancers in brightly colored costumes slowly moved along the main street. I got my camera and started taking pictures. After a while I realized that I was no longer paying attention to anyone except one very pretty girl, photographing her again and again. She noticed my attention, smiled, and said, "Gracias," when the whirling motion of the dance brought her close.

I followed her group to the end of the dancing route and introduced myself. The girl's name was Carmen; she came from a mining town in a high, remote corner of the Altiplano, two hours to the south by bus. She spoke very beautiful Spanish, slow, clear, and old-fashioned. Soon we were having dinner in a small restaurant, and then I had my first lucky break in Bolivia. The hostel owners celebrated the parade by turning hot water on for the first time in a week. Carmen had never taken a hot shower before.

The next day I invited her to join me on the trip to the rain forest. I've noticed over the years that the girls who like me are usually the adventurous type, so I never miss a chance to invite them on some risky escapade. Carmen had never been away from the arid highlands, and she really wanted to see the forests, but, of course, it wasn't an easy decision. I admired her bravery when she agreed. We went to a truck depot and got a lift for the two-day ride.

The road from La Paz to the Amazon Basin is promoted in tourist booklets as "the world's most dangerous road," and is popular among mountain bikers. It's not really that bad, just very narrow in one part where it's cut in an almost vertical slope. Nobody really knows how many people have died there. Most of them are impatient city folks on weekend excursions, although a few trucks and buses can be seen stuck in treetops below. If you drive carefully, there's nothing particularly dangerous.

Like most roads linking the High Andes with the lowlands on either side of the mountains, the Death Road is breathtakingly scenic. It climbs to a high pass just below the snow line before dipping into splendid cloud forests full of weird flowers and hummingbirds. The good thing about our driver was that he had a party the night before, so every few miles he'd stop for a nap, and we could explore the forests at different elevations—each time a different world. Eventually, he gave up and just let me drive. He was a funny guy with an endless collection of folk jokes. When I mentioned that I lived in the US, he said, "My brother works there! His name is Ernesto Salsa. Have you met him, by any chance?"

We spent the night camped near Coroico, a town standing on a hilltop and surrounded by scarlet-flowered *Erythrina* trees, with an old church and a wonderful view of the Andes. At dawn the driver lit a bonfire and made an offering of coca leaves to Pachamama, the Inca fertility goddess. This ritual has to be performed every morning in August. A few hours later we arrived in Rurrenabaque, the gateway to the Bolivian Amazon.

Rurre is a strange hybrid between a rugged frontier town and a cosmopolitan tourist center. Boatloads of gringos depart every morning on guided tours to jungle and savanna camps. After coming back the tourists relax and heal from insect bites in the town's few bars, mixing with Indians who come from remote villages to buy supplies, poachers waiting to meet pelt smugglers, loggers returning from vacations in their hometowns on the Altiplano, truck drivers recuperating after a trip from La Paz, and coca traders traveling their unadvertised routes.

For me this town was a treasure trove of information. I talked to hunters, national park rangers, ecotour guides, and former caiman-skin dealers. It required spending more time in bars and drinking more cheap cocktails than I had done in all my previous life. Everybody was very enthusiastic to share their knowledge, and although their tales often contradicted each other or were too tall to buy, certain facts emerged—and some couldn't be found in published literature. Now I knew, for example, that different species of caimans preferred to "sing" at different times of day. There were four caiman species in the area, but only one, the yacare caiman, was common.

One night we were sitting in a bar talking to a tour agency owner when a drunk local approached us and tried to chat Carmen up. When she tried to ignore him, he started groping her. International bar etiquette in such cases is strict: I took an empty beer bottle and knocked him on the head. His friends apologized for him and carried him away. I expected the incident to be over, but what happened next made things a bit more interesting.

In the morning, two soldiers showed up at our doorstep and informed me that their commanding officer—the guy who had groped Carmen—was challenging me to a duel. He was giving me the choice of weapons: AK-47 assault rifle or machete. They looked ashamed and were happy to leave when I told them I accepted the challenge and didn't care which weapon to use to relieve them of their commander. We were going on a forest trip that day, so the event was scheduled for the first morning after our return.

I talked to a few people in town and learned that duels were illegal and increasingly rare in Bolivia. It was considered totally inappropriate for military or police personnel to have duels with civilians. Challenging a foreigner was unheard of. The general opinion was that being killed wouldn't be the worst outcome for the officer, and that he was either still drunk or bluffing, never expecting me to accept the challenge. I made sure it was known in town that I was an expert in machete combat. It wasn't entirely true: I did study *kenjutsu* for two months once, but it's not the same thing. I also had to use a machete a lot on my first job in the US, but, of course, not for fighting—I worked in small-scale logging and tree care. I also mentioned a few times that all Russian boys learned the use of AK-47s in school from the second grade. That was a bit of exaggeration even for the Soviet times, but in my case it was true because I used to take extracurricular shooting classes.

Later that day we took a boat up a fast river to a jungle camp in the Andean foothills. It was in Madidi, one of the four national parks with the world's highest biodiversity. (The other three are Manú in Peru, Yasuni in Ecuador, and Pico da Neblina in Brazil.) Only a fellow naturalist can understand what an intense pleasure it is to escape to the rain forest after so many days of talking and drinking.

We spent a few wonderful days there, camping by a small creek. There were no people within at least ten miles, so we walked around naked—the only way to be really comfortable in the rain forest. High levels of biodiversity were very obvious: I remember looking into my belly button after a walk along a tapir trail and finding five species of ticks and mites inside. The only animals missing were the ones I was looking for. Every night I searched forest streams and ponds, but couldn't find any caimans. Once I saw a pair of red eyes that looked just right, but they belonged to a large aquatic rat. Apparently, rivers in the area were too fast for caimans. The only large reptiles I found were two snakes, a beautiful rainbow boa and a huge coral snake. I didn't know that coral snakes could grow to such size, so I assumed it to be a harmless mimic and somewhat carelessly picked up the angry snake by the tail. Of course, I didn't let it bite me (bites by nonvenomous snakes can be very painful and sometimes cause severe allergic reaction), but it was still unprofessional, especially for someone with thirty years of snake-handling experience. Later, I learned that it was a deadly Amazon coral snake, the largest of them all, and that the nearest hospital with appropriate antivenom was in Miami.

We returned to Rurre to arrange a trip to the savanna lowlands, where caimans were said to be much easier to find. I almost forgot about the duel, but as soon as we checked into a hotel, a soldier showed up and said that the officer was inviting us for a talk in the same bar.

To my great amusement, the officer came accompanied by his wife. He apologized, said the whole thing was a drunken joke, and talked a lot about his three charming kids. He offered to buy us cocktails. The moment he left to get them from the counter, his wife called him an idiot and a laughingstock of the town. Carmen kept quiet during the peace talks, but when we got back to the hostel I realized that she was extremely impressed and excited by the whole affair. Few girls get to have duels fought, or at least considered, over them nowadays.

The next morning we rode a motorized canoe downstream, into the long-deforested land of cattle-raising *haciendas* and seasonal lakes. Pink river dolphins frolicked in the water; huge jabiru storks watched us from trees. And there were lots of yellowish yacare caimans on riverbanks, some up to ten feet long but most much smaller. That river was a great source of income for Rurre because tourists were taken there on savanna tours. Nobody hunted in the area, and the caimans were very tame. I noticed that the pattern of black spots on their jaws was individual: all caimans had different spots. It was obvious that yacare caimans in the savanna would be easy to study, and being able to recognize each animal was an exciting opportunity, although I wasn't sure how exactly I could use it.

We got back to Rurre and embarked on a long journey through hot pastureland to a small nature reserve farther east. One lake there had the only easily accessible population of black caimans in Bolivia.

But when we rented horses and rode to the lake, we discovered that "accessible" wasn't really the case. The water level was so low that it was impossible to even see the water through the impenetrable wall of reeds and shrubs. As we rode back, disappointed, we spotted a foot-long yacare caiman walking toward the lake, probably from some dried-up pond far away. It was obviously exhausted, and had another half a mile to go. I decided to give it a ride to the reeds, but when I reached down and picked it up by the tail, it bit me. Its teeth were razor-sharp, and sliced my finger in a few places. That was the only time in my life I was bitten by a crocodilian. Of course, I took the baby to the lake anyway. Predator bites often take a long time to heal, but these cuts healed in just two days, probably because the caiman hadn't eaten for a long time and had few bacteria left on its teeth.

We hitched a car ride back to the La Paz-Rurre highway. The driver was a burly, unshaven, but very polite man in his fifties named Jesus. He invited us to his house in a roadside town. He owned an airplane and let me fly it a few times. I once took flying lessons in the US but never had a chance to get a license, so I seldom have an opportunity to practice. It was so nice to circle above the savanna in bright moonlight, with icy Andean peaks on one side and the endless expanse of the Amazon lowlands on the other. Jesus offered me the use of his plane for my future research. I couldn't wait to start it.

It was very cold in La Paz when we got there after a long day of riding on bananas in the back of a truck. Fresh snow covered surrounding hills. We went to the same hostel and holed up in our room, trying to avoid the backpacker crowd. I used to like backpacking, but in the last decade it has evolved from a risky pursuit by a small elite of explorers into a well-organized and popular pastime, and most backpackers are no longer interesting to talk to. All you hear are quotes from *Lonely Planet* guidebooks and phrases like "We already did the jungle and are considering doing Titicaca next." Thousands of young people, spoiled by well-developed infrastructure and abundant information, move between the same hostels, gringo bars, and designated attractions, seeing as little of the real country life as the rich folks on organized tours they despise. It used to be dangerous, guys. It used to be adventure.

It used to be real. I was less than forty years old, but I felt myself a bit of a dinosaur, a great traveler of the past, although, of course, I was a pathetic nothing compared to the truly great travelers like Richard Burton, Faxian, Ibn Battuta, or René Caillié, and I didn't even have my own *On the Origin of Species* to show for it.

I had a few days left in Bolivia, so I decided to rent a jeep and take Carmen to Salar de Uyuni. It's a very touristy place, but I figured that having our own wheels would let us get away from the crowds.

Salar de Uyuni is an immense salt lake twelve thousand feet above sea level. It fills with shallow water for only a month or two at the end of the rainy season. For the rest of the year it's an absolutely flat expanse of snow-white salt. Tourists mostly go to hotels built from salt bricks and to a beautiful island in the middle of the lake with a forest of giant cacti. But if you drive into remote bays surrounded by dry mountain ranges, you feel like you are alone on another planet. One of the best things to do there is to climb on the roof of the jeep and look into the sky. During the day, when it's relatively warm and you can lie naked after making love, the sky is dark blue. At night you need a warm sleeping bag because the temperature drops below freezing immediately after sunset, but the stars are so unbelievably bright that it feels like looking at a great city from an airplane. Before dawn, the great arcs of the Milky Way and the zodiacal light intersect in the zenith, and it gets really quiet—the only time of the day when there's no wind.

I drove Carmen to her little town. Carmen, who had never left home for more than a couple days before, couldn't wait to see her family and friends. I was sorry to leave her but expected to return in fourteen months.

I often wonder what it would be like to live my entire life in one place. I'd never be able to do that because it seems to take something I don't have. Or maybe I have something—a dispersal gene, perhaps—that makes it impossible.

Soon I was flying over the Amazon again, crossing the still-great forest on my way home. I was sure I'd be back next year to study yacares in savannas and black caimans in the deep forest.

But, of course, nothing went according to plan.

6 Honest Courtship

Crocodylus palustris

Where observation is concerned, chance favors only the prepared mind.
—Louis Pasteur

THERE WERE A FEW THINGS I had to do before continuing my research. Up to this point, all I did was simply observe. Now I had to get scientific.

First, I had to come up with a theory that could be tested, and run some initial tests to make sure it wasn't completely wrong. If it was, the sooner I gave it up and switched to another one the better.

Second, I had to develop a research plan. Sitting near a pond and trying to write down everything the animals do is not how scientific observation is conducted nowadays. You have to know exactly what you are looking for, and record your data in a way that minimizes subjectivity.

Third, I had to take care of my personal life. Li was in China, Carmen was in Bolivia, and being alone again was downright depressing. For a field biologist who has to travel so much, having a long-term relationship is never easy, so this was the most difficult task.

Well, at least loneliness stimulates scientific inquiry.

One thing about my alligators was particularly puzzling. They had two different displays—bellowing and headslapping—that seemed to serve the same purpose. Why have both? I saw only two possible answers: either their functions weren't really the same, or they worked best under different circumstances.

At first I was mostly thinking about the function. Displaying means telling others something about yourself. What could each of the two displays tell about the animal making it? The frequency of bellowing sound depends on the size of the alligator. Headslaps don't contain that information, but they have another advantage: it's very easy to tell which direction they come from. Many animals, including humans, locate the source of sound using the tiny difference in the time of arrival of sound waves to their left and right ears. This difference is most easy to detect if the sound has a sharp onset, like a gunshot, a crack of a whip—or a slap.

So bellows can tell you how big the animal is, and slaps can tell you where it's located. It seems logical for the alligators to have just one type of display serving both purposes: a bellow followed by a headslap. "I'm a big strong male, and I'm in this pool. Ladies welcome." But that's not what is happening. Male alligators use either bellows combined with infrasound, or slaps combined with infrasound. Why?

An obvious way to study the function of these signals would be to use playbacks: just play some recordings to alligators and see how they respond. But I soon learned that playing infrasound underwater would require a lot of energy and some heavy equipment. Such equipment could be rented from the navy, but it would be mounted on a big truck, and monthly rent would be so high that only a very large research grant would cover it. No wonder the only other animals known to make infrasound underwater are big whales.

Wow, I thought. Six-foot-long alligators do something that people need a truck to reproduce. Why would these somewhat phlegmatic creatures spend so much energy?

One concept popular in modern ethology is *honest signal*. During courtship or conflict, animals (including ourselves) always try to make a good impression. And they often cheat to seem more attractive or more fearsome. Women use breast implants and high heels; lions grow manes to look bigger; men have their larynges descend in their throats at the time of puberty to make their voices sound more masculine.

(Interestingly, male elk use the same trick, presumably to sound as if they were larger.) The recipients of these signals, be it the opposite sex or the rivals, always try to evolve the ability to see through the cheating. In many cases, the species eventually settles on using signals that can't be cheated. Songbirds sing very fast songs with crazy jumps in pitch, and doing so is extremely difficult and requires top fitness. Hummingbirds perform elaborate displays that require outstanding mastery of the air. Bighorn sheep have duels that directly test their physical strength. Some women believe that money is the most reliable measurement of a man's fitness (I can only hope they are wrong).

What if alligators use infrasound as an honest signal? Producing it takes immense physical strength, and, as I found by trying to fake it, is cheat-proof.

Infrasound can carry for great distances through the water; whales use it to communicate over hundreds of miles. Locating the source of infrasound underwater is very difficult for a number of physical reasons, but accompanying it with headslaps solves the problem. So a combination of infrasound and a slap is a perfect way to introduce yourself and to give directions for those interested.

OK, but what are bellows for?

I watched my video recordings of bellowing alligators and saw that, unlike infrasound, which is produced by the submerged part of an alligator's body, bellows are emitted above the water surface. Maybe they are supposed to carry information about the animal through the air, while infrasound and headslaps carry it through the water?

This was a neat idea, but it still didn't explain a lot of things. Why do females produce bellows and headslaps, but no infrasound? What is the purpose of bellowing choruses? I decided to return to these questions later. For now, I wanted to test one theory: that each type of display was optimized for a certain medium. And I immediately thought of a simple way to test it.

If bellows are for the air, and slaps are for the water, then animals living in small ponds should use more bellows. Making slaps there doesn't make much sense because the underwater signal wouldn't get very far. But animals living, for example, in large rivers should use more slaps, because underwater signals can carry much farther provided there's enough water.

That was still far from having a well-formulated theory and a real research plan, but at least I had a testable prediction. A nice chance to test it was coming up during the winter break. Two of the three crocodilian species living in India were supposed to mate at that time of year. So I prepared for a field trip.

The two species I was about to meet were the mugger crocodile and the Indian gharial. The mugger, also known as the marsh crocodile, is much more common, occurring in numerous nature reserves in India and neighboring countries. Surprisingly, for such a widespread species, there was no published data on its courtship behavior, except one unlikely source. In *The Undertakers*, one of Rudyard Kipling's tales from *The Jungle Book*, there's a beautiful description of the mugger's impressive roar. I later used it as the epigraph for my thesis:

> "Respect the aged!"
>
> It was a thick voice—a muddy voice that would have made you shudder—a voice like something soft breaking in two. There was a quaver in it, a croak and a whine.
>
> "Respect the aged! O Companions of the River—respect the aged!"

Muggers can live almost anywhere, from small ponds to coastal mangroves. So in theory I could compare mugger "songs" from two habitats—but that would take much longer than one month-long winter break, because I'd need a large sample size to prove any differences. The best I could hope to achieve in one month was to get basic information about their displays and see if they generally match my predictions.

I wrote emails to Indian crocodile researchers. Father and son Romulus and Nikhil Whitaker, a famous team who have done more than anyone to save Indian crocodiles and gharials from extinction by captive-breeding them, were very helpful. They told me that muggers were easy to find, but gharials existed in the wild in only a handful of places, with the largest populations in the Chambal and Katarniaghat nature reserves. Chambal was less than two hundred miles southeast of Delhi, but it had gharials scattered along a hundred miles of the river and was considered unsafe to visit. All over India this place was known as the favorite hideout of *daikoti* (professional bandits). Katarniaghat had a dense gharial population but was very remote. I decided to start from Corbett National Park, a place with both gharials and muggers within a day-long bus ride from Delhi.

One week before my departure I got a phone call from a Russian named Boris who was a big fan of my travel books. He was coming to Orlando for a seminar on something banking-related and asked if he could visit me in Miami afterward. I invited him to join me for an evening walk in the Everglades.

A few days later he showed up at my door and said:

"There's a girl waiting in the car. She attended the same seminar and gave me a ride to Miami. May I invite her for a cup of coffee?"

"Sure," I replied, expecting to see an archetypal bespectacled middle-aged bank accountant. But Boris returned with a creature so beautiful that she looked like a heavenly apparition in my modest Little Havana apartment. She was very young and slender, spoke with a charming West Indian accent (I later learned that she had grown up in Jamaica)—and was a bank accountant, although a high-ranking one. Her name was Kami.

Well, Boris expected to be taken to the Everglades, and I could think of nothing better than to invite Kami to join us and check out some alligators.

"I've never been to the Everglades and never seen an alligator," she said.

"How long have you lived in Florida?"

"Twelve years."

It took me a while to realize that she wasn't joking. I was about to enter the strangest relationship of my life. It was like falling in love with a space alien.

Kami was very different from the kinds of girls who usually found me interesting. Most of those girls were the ones who wanted to be pirate captains when they were little. When Kami was little, she wanted to be a bank CEO. She later asked me never to mention that evening to her friends because it was totally out of her character to agree to something as scary as a night walk to a swamp.

If you look at a map of Miami, you can see that the western border of the city is almost a straight line, a street called Krome Avenue that separates continuous urban areas along the coast from the uninhabited expanse of the Everglades in the interior. Kami lived east of that divide, never bothering to cross it. I also lived in the city but tried to spend as much time as I could on the wild side. In sixteen months I hardly ever visited downtown Miami or the civilized, hotel-lined urban beaches. If I wanted to swim in the ocean, I'd go to more remote shores farther south. Kami and I lived just a mile apart, but we existed in completely different worlds.

It was getting dark already, so I took Kami and Boris for a short walk in Shark Valley, one of the most touristy places in the Everglades. Normally, it has visitors even late at night because many locals enjoy walking or biking the long paved trail across the marshes. But this was a cold, windy, moonless night, and the three of us were completely alone. I kept telling Kami that the trail was perfectly safe, but she was horrified. One particularly loud splash in the swamp made her jump into my arms. I was cursing myself the whole time, certain that I'd never see her again. I was wrong.

We started exchanging phone calls and even had a formal date in a restaurant just before my flight to New Delhi. I had to wait six weeks for the next chance to meet her, but the emails I wrote when I was away would make a decent-sized volume of sensual poetry. Most of them were rather sad: in addition to being painfully lonely, I ran into serious problems in my work.

I landed in Delhi at midnight and got out of the city at dawn, but that wasn't fast enough to avoid the respiratory infection that almost all first-time visitors get there because of the severe air pollution. Like most big Indian cities, Delhi is very intense. The air is so thick you can barely see as far as a few blocks, the crowds on sidewalks often swirl into huge traffic jams, and vehicular traffic is the world's most challenging driving experience (there are signs saying MAXIMUM SPEED ON THE

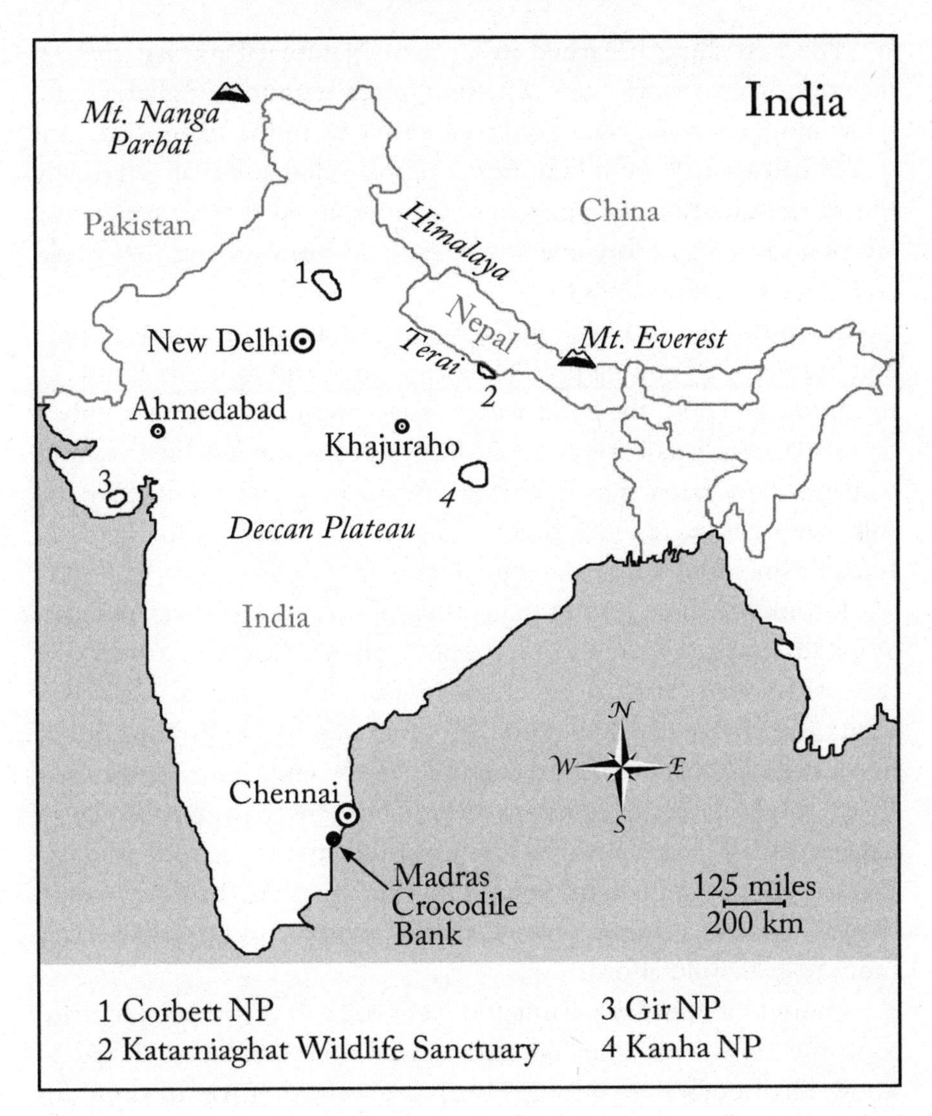

Opposite Side Of Road 100 km/h). Traveling around with a fever wasn't easy. Nights were bitterly cold, and I didn't have winter clothes because I hadn't expected such low temperatures in tropical lowlands. Sometimes I had to walk around wrapped in my sleeping bag —but locals didn't mind because it looked a bit like the traditional Hindu way of dressing that you have certainly seen in Gandhi photos.

The first thing I learned in Corbett National Park was that I'd never be able to work there. The only place with lots of gharials and a few muggers was a river pool well visible from the main road, so it would be easy to watch them from the roadside—but that was totally out of the question. The park staff were so afraid that a tiger would attack a visitor that foreigners were banned from leaving their jeeps and fenced campgrounds.

So I returned to Delhi and traveled east along the Terai—a long belt of lowlands running parallel to the Himalayan foothills. Until the invention of DDT, the Terai was sparsely populated, because only a few malaria-resistant tribes could live there. In the last fifty years its wetlands have been mostly converted to rice paddies, but there are still many nature reserves there, some of them famous for herds of Indian rhinos and wild Asian buffaloes.

It took me three days of train-hopping to get to the Katarniaghat Wildlife Sanctuary on the Nepalese border. There I was told that research permits had to be obtained from the regional office, six hours away by a very slow train. The office was in a crumbling building that clearly hadn't been repaired since the end of the British Raj. It was filled up to the ceiling with moldy, cobweb-wrapped piles of paperwork. Getting a research permit would take a year, they said, and I'd have to stay in Delhi to keep things moving. The best they could do was issue me a tourist permit, and that would cost $80 a day—way more than I could afford.

I bought a one-day permit and went back to the reserve. A guard took me on a boat tour along the river and told me all I had to know. The books were wrong: North Indian winter was too cold for gharials and crocodiles to mate, so they wouldn't start "singing" until March. I was welcome to come back in spring without any permits and stay in the guard's family hut at the edge of the reserve for a dollar a day, meals included.

He explained that the reserve border wasn't clearly demarcated, so there were lots of riverbanks and forest ponds I could walk to without breaking the law. Technically, I'd still need a permit for any research in India, but only if it wasn't tourism.

"Is your research tourism?" he asked.

A strange question, but my research was purely observational. I didn't interfere with the animals' behavior in any way. I said it could be considered tourism.

So we agreed to meet again in March, during my spring break. In the meantime, I decided to move farther south in search of warmer weather.

Some of the best national parks in India are clustered in the central state of Madhya Pradesh, east of the ancient temple complex of Khajuraho, famous for its thousands of erotic bas-reliefs. These parks had plenty of mugger crocodiles, but again it was impossible to get permission to enter them on foot, and hiring a jeep was too expensive. Park administrations were totally paranoid about tigers, even in smaller reserves that had only one or two left. Interestingly, locals were much more afraid of leopards and sloth bears, and for a good reason. These animals kill a lot more people in India than tigers do. Leopards are particularly sneaky and deadly. Recently, they took a few residents of Mumbai, a city of twenty million.

Eventually, I found a little forest pond located precisely on the border of Kanha National Park, set up my tent on the shore, and started watching muggers there. The pond contained two six-foot crocodiles.

The next morning, after the sun came up and it got a bit warmer, I decided to walk along a forest road. There were fresh tiger footprints in the sand, so I got my camera ready. The road made a sharp turn, and there she was—a large tigress, resting on the shoulder.

She looked at me, and I instantly felt uncomfortable. Anyone who's ever had a house cat would recognize that focused gaze, as if there was a fly sitting on my forehead. Still keeping my eyes on her, I slowly took a few steps backwards, toward a large sal tree. But before I managed to get behind it, the tigress charged.

I flew up the tree like a squirrel, but the highest I could get was about twenty feet. The tigress paused for a few seconds and started cautiously climbing after me. When she was almost within a paw's reach, I put away my camera, broke off a large branch, and poked her in the nose. The tigress hissed, and I realized that if she moved up another inch, my only choice would be to hit her in the eye. I poked her in the nose again and again, she tried to give me a slap, but her claws slid, and she clumsily descended, almost falling to the ground.

She circled the tree once and walked off. I remained in place, waiting patiently. Five minutes later, a shrub about two hundred feet away moved, and I saw her watching me through its foliage. Her face looked so cunning that I couldn't help laughing. Realizing that she'd been spotted, the tigress yawned, turned around, and disappeared. Soon I heard a loud whistle—an alarm call of a chital deer—from at least a quarter of a mile away. Now I could be sure that she had left. I got off the tree, looked through the pictures in my camera, and started laughing again. The last photo was shot from such a close range that you could see a big fat tick attached to her cheek.

Unfortunately, Kanha nights were still too cold, and the crocodiles barely emerged from their deep burrows. I was getting really worried: a second expensive overseas trip was about to become a total waste of time. So I packed up, and after two more days of train-hopping arrived in Madras Crocodile Bank in South India.

This "bank" is not really a bank, of course. It was started by Romulus Whitaker in the 1970s as a breeding facility, and helped repopulate many of the country's nature reserves with crocodiles and gharials. Nowadays captive breeding is no longer important, because the main problem for Indian reptiles is loss of habitat rather than hunting. So the bank gradually evolved into an excellent research facility. It's also a large reptile zoo, with hundreds of muggers, among other things.

When you study animal behavior in the wild, it's always helpful to watch the same species in a zoo. There are lots of things that are very difficult or impossible to observe in any other way. So I spent the first night and morning of the new year watching muggers in the bank's largest enclosure. The night wasn't a merry one: the New Year in Russia is traditionally a family holiday, and I was there alone, without even the Internet to chat with Kami. Nothing interesting happened all night except for an egret chick falling from a tree (it was swiftly caught and swallowed by one of the crocs). Shortly after dawn, the largest male roared—gave a loud, coarse sound that Kipling had described so well. Three nearby females immediately ran up to the male and put their heads on his back. I already knew from observing alligators that this was common foreplay behavior.

Nikhil, son of Romulus Whitaker, who is also a famous herpetologist, suggested that I try Gir National Park in West India. It's near the coast of the Arabian Sea, so winters there are relatively mild. The park is famous for having the last surviving population of Asian lions, but it has plenty of other wildlife, including a few hundred muggers. I crossed the country to its western coast, and hitched a two-day-long truck ride to Ahmedabad up north. The driver smoked hashish the entire time, so I had to do most of the driving.

When I finally got to Gir it was late night. The village was sleeping, so I walked into the forest to camp. West India is generally very dry in winter; the night was warm, and I was very tired, so I decided to sleep without the tent, choosing a place under a teak tree. Its huge leaves form a thick layer underneath that makes a lot of noise if anything, even a small beetle, walks on them—so it would be impossible for any predator to approach me unnoticed.

At dawn I was, indeed, awakened by a predator, but not by the soft steps of a stalking leopard. A lion was roaring not more than three hundred feet away. It took me a few minutes to master all my courage, crawl out of the sleeping bag, get my camera, and walk toward the sound. I kept telling myself that Gir lions were generally considered safe for humans, especially compared with local leopards—those had a really bad track record.

Instead of walking into a lion, I was stopped by a dilapidated brick wall. The sound was coming from the other side. With considerable effort, I climbed the wall, and finally saw the roaring beast. It was in a cage.

Local people walk freely around the park, but, as usual, foreigners are banned from entering on foot. I talked to a park ranger, and he promised to show me a patch of forest outside the park with plenty of muggers. He charged me a few bucks for the favor, of course. Only later did I find out that he had led me straight into the heart of Gir, and that I'd be in a lot of trouble if other rangers saw me. I spent a week camping there, sleeping during the day and watching muggers from dusk till late morning. It was difficult not to get distracted by lions, leopards, antelopes, and deer walking by, but I finally logged some data on muggers.

Gir National Park is covered with dry tropical forest. It can get very wet during the monsoon season, but in the winter and spring months it's so dry that most trees drop their leaves. This is what is properly called *jungle* (a Sanskrit word meaning "dry land"), although since the time of Kipling the term has been used for any dense tropical growth. In early January only a few forest streams still had water. Muggers were stuck in small river pools, separated by miles of shallow creeks barely trickling along their gravel beds. Every morning, just before or after dawn, the largest crocodile (presumably the dominant male) in each pool would roar or headslap. There was nothing like alligators' bellowing choruses; all other crocodiles were completely silent, except for a few hisses and growls given in aggressive encounters. Their roars sounded a bit like those of male lions, and once I saw a mugger roar and produce infrasound in response to a lion roaring nearby. It was a beautiful exchange—I wish I could have recorded it on tape, but it happened too suddenly.

The crocodiles remained active at night even when it was chilly, and big ones would sometimes crawl out of the water to spend a few hours lying still at the edge of a riverside trail. I remembered seeing large alligators doing the same thing at night, but why? Here was something to think about.

As I counted my observations on a train to Delhi, I was glad to see that there were many more roars than headslaps, just as my theory had predicted for crocodiles living in small, isolated ponds. So this idea was still worth working on. However, I have to admit that on my way home I was mostly thinking about my own courtship strategy.

7
Hunting on Shore
Crocodylus acutus

Loving someone like you is easier; loving someone different is more fun.
—Desmond Tutu

BY MID-JANUARY MY LIFE WAS PERFECT. I had a beautiful and intelligent girlfriend, developed a draft theory to work on, passed my qualifying exams, and even managed to obtain a small research grant, although it barely covered the airfare to India. Every day was a discovery—not in zoology (alligators don't move much in mid-winter) but in everyday life, thanks to Kami.

She'd take me to places I'd never even considered visiting: nightclubs, spa salons, expensive restaurants. I also tried to introduce her to new things by taking her on short hikes or kayaking trips in the Everglades. For both of us, one of the most interesting parts of this experience was the other's race. Kami had never had a white boyfriend before. I was once in love with a dark-skinned village girl on Madagascar, but I was there on a tourist visa, and barely had time to buy her out of slavery, teach her a little zoology, and find her a job as a tour guide in a nature reserve. In the US I always lived in places with relatively few African Americans: California, Colorado, New Mexico. So the lives of that segment of the country's population were mostly a mystery to me.

Like many African American women in the US, Kami straightened her hair. I found this practice conformist and almost as perverted as the barbaric custom of bleaching one's hair with peroxide that some white tribes still practice. Kami had a perfectly shaped head and neck. Why cover them with hanging hair? I asked her if she had ever considered "going natural." She couldn't believe I was serious. Apparently, many black girls think that "African" hairdos are totally inappropriate in business environments and would ruin any mixed-race relationship.

It took me almost a month to persuade Kami to change her hairstyle. I even cut my own hair to almost nothing as a sign of support. But on the day she finally decided to do it, I was teaching at the university and couldn't accompany her. So the poor girl had to endure the ordeal on her own. And it was, indeed, an ordeal. Three hair salons refused to cut her hair. Perhaps they thought she was crazy or going through some personal tragedy, and didn't want her to run them through with scissors or to commit suicide on their doorsteps. By the time I called her, she was getting desperate.

"Just go to the next one and tell them you are moving to Uganda," I said. That worked. When I met her in the evening, I was rewarded for my persistence with a wonderful sight: Kami looked like a delicate forest nymph of African fairy tales. Touching her head was pure pleasure. I just couldn't get my eyes—and hands—off her.

Kami was still worried. What will people at work say?

"How did it go?" I asked her, calling the next day during her lunch break.

"You know, they liked it! I have a feeling that they all wanted to touch my head but were afraid to ask!"

Having an interracial relationship wasn't always easy. In February we went to the Dominican Republic for a few days in an attempt to combine a beach vacation with crocodile observations. Everybody in that small country automatically assumed Kami to be a local prostitute, and me her client. The fact that she didn't speak Spanish made things even worse, because she was suspected of being a Haitian refugee. At first we tried to laugh it off, but after a few evenings spent in fruitless attempts to check into hotels it started driving us mad. We ended up camping in remote woodlands most of the time. It was fine with me (my idea of a good beach is one with no people within at least a mile), but Kami thought it was way too risky.

She also thought that I was too risky for her in general. No matter how hard I tried to behave normally, she still considered me wild and unpredictable, and refused to discuss our relationship as something more serious than a short-term adventure.

And while my personal life was getting into trouble, my research was threatening to become a total mess.

Crocodiles of the Dominican Republic belong to a species called the American crocodile, which prefers brackish water and inhabits mangrove lagoons from Mexico and the West Indies to Venezuela and coastal Peru. Those in the Dominican Republic live in a huge salt lake below sea level; the salinity there is so high that the reptiles stay in places where small rivers flow in.

The northernmost population of the American crocodile inhabits the southern tip of Florida, making it the only part of the world to have both crocodiles and alligators. Florida crocs have recently been taken off the endangered species list, but there are still less than two thousand adults, most of them living in the cooling canals of the Turkey Point nuclear power plant near Homestead. The only place in the States where they are easy to see in the wild is Flamingo, a ranger station and fishing camp at the end of the long road crossing the southern Everglades. If you get there in the morning in late February or early March, you'll probably hear the dominant male roar, or, more likely, headslap. And it would probably be one of the males I've recorded. I spent a lot of time there, paddling in my kayak along the shore and into the endless dark labyrinth of mangrove channels, trying to see displays by as many crocodiles as possible.

American crocodiles got me thinking about a second method of testing my hypothesis that roars and bellows were for communication through the air, while headslaps were for communication through the water.

Most crocodilians are *habitat generalists*: they can inhabit virtually any available body of water. Muggers and American alligators are

The Caribbean
300 miles
480 km
USA
Miami
Havana
Cuba
Haiti
Dominican Republic
Puerto Rico
Santo Domingo
Jamaica
Mexico
G
H
N
Managua
T&T
Caracas
Venezuela
Orinoco
Colombia
Bogotá
N
W
E
S
DR Dominican Republic
G Guatemala
H Honduras
N Nicaragua
T&T Trinidad & Tobago
1 Zapata Swamp
2 Lago Enriquillo
3 Tuxtla Gutiérres
4 Selva Lacandona
5 Sierra de los Cuchumatanes
6 Útila Island
7 Corn Islands
8 Lago Nicaragua
9 Riohacha
10 Tayrona NP
11 Salamanca NP
12 Barranquilla
13 Cartagena de Indias
14 Moroccoy NP
15 Tucupita
16 Sierra Nevada de Santa Marta
17 Ciudad Bolívar
18 Cordillera de Mérida
19 Hato Masaguaral
20 Rio Capanaparo
21 Puerto Ayacucho
22 Mt. Roraima
23 Auyán Tepui
24 Autana Tepui
25 Santa Elena de Uairén
26 Black River Great Morass
27 Blue Mountains
28 Cockpit Country
Jamaica
Kingston
50 miles
80 km

good examples. The first way to test my hypothesis was to compare displays of habitat generalists living in small ponds and in large lakes or rivers.

The remaining species are *habitat specialists*: they live only in coastal habitats (like the American crocodile), or only in large rivers, or only in forest swamps. So instead of comparing populations I could compare species. Chinese alligators, for example, live only in relatively small lakes and ponds. They use a lot of bellows, but very few headslaps. I was happy to find that American crocodiles in both the Dominican Republic and Florida used a lot of headslaps, but seldom roared.

The problem with this approach was that there were only five species confined to small bodies of water, and five species confined to large ones. With such a tiny sample size, it would be extremely difficult to get a small P-value.

If there's one thing you need to know to understand how statistics are used (and sometimes misused) in biological and medical studies, it's the concept of P-value. This number is so important that a popular toast among biologists is: "Let your list of publications be huge and your P-values small."

"P" stands for probability. Imagine that you throw a die twelve times and get five sixes. Is the die rigged, or was it just a random event? There is a statistical formula that you can apply to find the probability of a "normal" die giving you sixes five times out of twelve. This probability, or P-value, is 0.02 (two out of a hundred). You still don't know for sure whether the die is rigged or not, but it seems very likely that it is.

If you are comparing data from two populations, there will always be some difference in numbers. Again, P-value is the probability that the difference you observe is just a random fluke. The smaller the P-value, the more likely it is that the two populations are really different.

If that real difference does exist, getting a larger sample size would give you a smaller P-value and more certainty. If it doesn't, you can increase the sample size all you want, but the P-value would only go up. So the larger your sample size, the more certain the answers you are likely to get.

Of course, all scientists would like their P-values to be zero, so that their answers would be 100 percent certain. But it seldom happens in real life. In biological research, a P-value of less than 0.05 is usually considered convincing enough. If you get below 0.05, your result is called *statistically significant*, but that doesn't mean that your theory is correct—it's just very likely to be correct. On the other hand, if you can't get below 0.05, it doesn't mean that you are wrong—just that you failed to obtain sufficient proof. So if you read in a newspaper that, for example, "a recent study found no connection between cell phones and brain cancer," it doesn't necessarily mean that cell phones are absolutely safe. Maybe they do cause brain cancer, but so rarely that the researchers didn't have enough data to prove statistical significance.

Comparing various crocodilian species was unlikely to get me a small P-value, but I still had to do at least a basic study of as many of them as possible. My theory predicted that their use of roars/bellows and headslaps depended on their habitat preferences. I didn't want to spend years working on it just to find out a week after publishing the results that some little-known species' behavior completely contradicted that prediction.

As for comparing animals of the same species living in different habitats, I could get a much larger sample size, but that would take years, because the mating season of each species was only a few weeks long.

There was no way around it. I needed a small P-value.

But what bothered me the most about my theory was the ever-growing list of things that it couldn't explain. For example, alligators used either bellows or headslaps, while crocodiles could combine roars and headslaps in the same display. I couldn't come up with any possible explanation for this difference.

Meanwhile the crocodiles' mating time was over. As the dry season progressed and less fresh water was entering the Everglades from the north, brackish water was slowly moving inland, and so were the crocodiles. They appeared in lakes previously inhabited by alligators. Being more aggressive, crocodiles usually became dominant with little or no fighting as long as they and the alligators were of similar size. Soon alligators left those lakes altogether. Unlike crocodiles, they can live in brackish water for only short periods of time, because they lack special glands in their eye sockets and tongues that excrete excessive salt, producing the infamous "crocodile tears."

One night I was walking along a lakeshore and saw a crocodile ahead. Its eye reflected my flashlight and looked like a hot coal. The croc was lying near the trail with just its snout sticking out of the grass. It was almost ten feet long—crocodiles in Florida don't grow much larger than that. I remembered seeing adult muggers and alligators doing the same thing, sometimes up to a hundred feet from the water. In the touristy Shark Valley you could often see dozens of alligators lying for hours on the road shoulder, always at night. What were they doing?

I thought they were in ambush, waiting for some animal to walk by. I discussed it with other crocodile researchers, but they were skeptical. The established dogma was that crocodilians were aquatic hunters and virtually never hunted on land. The alternative explanation was that the reptiles' behavior had something to do with thermoregulation: perhaps they were attracted to warm road surfaces. I got a laser thermometer and found that just two hours after sunset the road wasn't warmer than the grass or the air anymore, and that it was always colder than the water from which the alligators or crocodiles were coming. Then someone suggested that the animals were trying to dry their skin: they need to do so from time to time to prevent it from getting overgrown with algae. But why do this at night, when there's dew everywhere, and not in the sun?

I became convinced that my explanation was correct when I saw an alligator with a recently caught opossum in its jaws on a trailside. But I knew that I needed more observations to prove my case.

I went through the records of alligator attacks on humans in the United States and found that a few lethal attacks had occurred on trails up to thirty feet from water. I wrote a paper about my findings, but decided to wait for a while before submitting it to a journal, just in case I'd get more data. I eventually had it published three years later, by which time it included observations of trailside ambushes and hunting on land by three other large species: the Nile crocodile, the saltwater crocodile, and the black caiman. But for now, I was the only person in Florida who knew how dangerous it was to walk on trails along rivers or canals at night without a flashlight.

Living in Florida means that you get a lot of visitors. As soon as I moved to Miami, old friends began showing up, especially in the winter and spring months. Once I took two weekend visitors from New York to the Everglades, and we decided to walk Snake Bight Trail. In the summer this trail has the highest density of biting insects in the world. A biker once died there after the chain of his bike broke at the far end of the trail; poisoning by mosquito saliva killed him before he could walk his bicycle three miles back to the highway. But in the winter there are hardly any mosquitoes or blackflies there, and it's a nice easy walk through the woods.

By the time we were walking back it was already dark. Suddenly we saw a nine-foot alligator lying across the trail, completely blocking it. We had no choice but to jump over the reptile. I shone my flashlight directly into its eyes, blinding it, while my friends leaped over its back. The alligator didn't move a muscle: it was still too small to feel confident hunting humans. Alligators (and crocodiles) don't normally consider people as prey until they grow to around twelve feet. Very few alligators in Florida ever grow that big, and that probably explains why attacks on people are relatively rare. In a state with nineteen million people and one million gators, less than twenty people have been killed by alligators in the last hundred years. As for crocodiles, they have never killed anyone in the US, although the same species does sometimes kill people in other countries, where the crocs can grow much larger.

By mid-March I realized that I had an enormous amount of work to accomplish. I had to study mating-season displays of as many crocodilian species as possible, and I had to collect a huge amount of data on American alligators—both of which would require spending a lot of time in the field. But I also had to teach for at least eight months a year to get my stipend and avoid paying tuition. The project could easily take eight or ten years, while I was supposed to defend my thesis in four.

I didn't have enough time for field research, and I also ran into totally unexpected trouble with teaching. My job was to teach introductory biology labs. Many of my students had not chosen a career in biology; they just needed it as a prerequisite for medical school. These kids cared little about the subject, but they worried obsessively about getting A's, and if they couldn't, they'd sometimes try to force me by complaining about something to the department head.

One of the labs had to do with various kinds of segmented worms, such as leeches and earthworms. I had to find a volunteer to demonstrate the feeding behavior of a medicinal leech by letting it latch on to the person's hand. (Interestingly, the volunteers were usually girls.) I also had to explain the basic facts of worm biology. Earthworms are hermaphrodites; when they mate, it goes both ways. The next week there was a quiz, and I included a question on earthworm reproduction. I went to Wikipedia, found a list of sexual positions, gave it to the students, and asked them to choose the one that earthworms were most likely to use. Anyone who had listened during the previous lab or read the textbook would know that the position had to be symmetrical, so the only possible answer was sixty-nine. Almost everybody answered the question correctly, but one girl whose grades were far from A's decided to write a complaint.

So I had to go to the office of our department chair and listen to a lecture about the inappropriateness of sexual innuendo in quizzes. Poor Steve, my advisor, had to sit there with me, and I could see what a heroic effort it was for him to not burst into laughter. I don't think anyone has tried teaching biology without sexual innuendo since the late eighteenth century, when Catholic schools banned teaching Linnaeus's botanical classification because it was based on differences in flower structure (flowers are sexual organs).

Meanwhile, Kami said that she was breaking up with me. It was a strange breakup, because we'd still meet two to three times a week. But I knew that it would be over as soon as she found some other option. For such a beautiful girl it wouldn't take long to find one.

And now I had to leave her for two long weeks to go back to India.

8
Companions of the River

Gavialis gangeticus

The first condition of understanding a foreign country is to smell it.
—Rudyard Kipling

YOUR FIRST DAY IN INDIA IS USUALLY A NEGATIVE EXPERIENCE, even if you have been there before and know what to expect. My trip to China the previous year was still fresh in my memory, and the comparison wasn't in favor of democracy (which India is). While China's one-child policy has resulted in an unprecedented economic boom and, recently, some impressive conservation successes, India has failed to implement anything of the kind. Political parties that tried to curb the birthrate have invariably been voted out of power. The resulting population explosion has left the country without any hope of quality-of-life improvement for the majority of its people.

Despite being so colorful and photogenic, India can get really boring. Its transportation network is outdated and inadequate, so traveling a hundred kilometers by bus or narrow-gauge railway sometimes takes a whole day. Lifting a road barrier after a train has passed can take up to three officials and twenty minutes. In South India most people speak at least some English, but in the northern part the lingua franca is Hindi. Unless you speak Hindi, your interactions with other passengers are more like those of a zoo animal with the crowds gathered around its cage. If someone who speaks English shows up, the situation often reverses and you can't get out of an endless chat with dozens of people. Well, at least it's considered inappropriate to bug you with religious questions. In Pakistan almost every conversation starts with "Are you Muslim?" In India I've been asked that only once, in the Muslim quarter of Jaisalmer.

My three-day train journey to the Katarniaghat Wildlife Sanctuary was particularly uneventful. For almost every visitor to India, the main source of surprising, unusual, and funny adventures is traveler's diarrhea, which I, having already introduced a lot of tropical microbes to my system, was spared. So my only entertainment was watching endless cities and towns pass by, helping people who came down with heatstroke, and assisting a tiny low-caste girl in delivering a baby (no, her family didn't show any intention of naming the boy after me—they were almost dead with embarrassment).

The two largest religious communities in India are Hindus and Muslims. Watching local wildlife from a train can often tell you which of the two is living nearby. In Hindu-populated areas, the only large animal still common outside nature reserves is the nilgai, an antelope that resembles a cow and so is not eaten by Hindus. In places where Muslims are the majority, the most common large animals are wild boars and porcupines, because the meat of both is considered pork.

The park ranger who had invited me to stay with his family for free just two months earlier now wanted a lot of money for his hospitality. I set up my tent in the forest outside the nature reserve. The forest was heavily used for gathering firewood and grazing cattle, so during the day a lot of people walked by my camp, but nothing ever got stolen. At night and early in the morning there were other visitors: sloth bears, striped hyenas, golden jackals, langur monkeys, beautifully spotted chital deer, and large, heavy sambar deer. The deer and the langurs were my early warning system. I knew they'd sound the alarm if a tiger or a leopard was in the area. But I never saw any cats, only their tracks in the mud near forest ponds where I watched mugger crocodiles.

The river, oxbow lakes, and even small ponds had lots of muggers. If the pond was about to dry out, the crocodiles would excavate long tunnels in its banks and hide there. As the days got hotter, crocodiles that lived in ponds and lakes disappeared in their burrows almost completely, showing up only late at night to wait on forest trails for passing mammals or to dig in the bottom mud for snails, clams, crabs, and turtles. Their massive jaws were well suited for cracking shells. Muggers that lived in the river spent hot days on sandy shallows, and I could sometimes get really close to them in my kayak. The largest mugger in the area was a male almost twelve feet long, with a head so huge and scary that I nicknamed him T. rex. Later I learned that locals called him Surasa after a sea demon from the Mahabharata, an ancient Indian epic.

In Katarniaghat, where many muggers lived in a large river, the proportion of headslaps in their "songs" was higher than in Gir National Park, where they all lived in small forest ponds. That matched my predictions, but my sample size wasn't large enough to give a good P-value. I couldn't get more observations because muggers usually headslapped or roared only once per day, and the mating season was almost over.

Muggers were interesting to watch, but it was the other giant reptile of Katarniaghat that I found totally captivating.

The Indian gharial is a living fossil. Of course, any crocodilian can be called a living fossil, but this one qualifies even more than others, and it looks the part. The last survivor of a separate lineage that used to exist on all continents, it now lives in just a few remote areas of North India and Nepal. This specialized fish-eater looks like the exact opposite of the mugger. Its jaws are almost unnaturally long and thin, with comb-like rows of long, delicate teeth. Unlike muggers, which kill fifteen to twenty people a year, gharials are not known to kill humans—although they do eat corpses.

Smooth-scaled and long-tailed, gharials are excellent swimmers: I saw them fishing in rapids side by side with river dolphins. But, unlike other crocodilians existing today, gharials cannot walk on land, only crawl. And unlike muggers they are very shy. I had to observe them from shore because they wouldn't let me get close in my kayak.

Adult males were about fifteen feet long and had a huge, tumor-like growth at the tip of their snout. The word "gharial" comes from *ghara*, a Hindi word for "cooking pot"—apparently, that's what this enormous organ resembles to local people. Its function is still a matter of controversy.

Males with gharas spend hours lying on the beaches with their heads lifted up so that the ghara can be seen from a mile away. Like some researchers before me, I suspect that this posture is a signal, attracting females and warning other males.

Other crocodilians don't have gharas, and you cannot tell their males from females by the way they look, unless they are too big to be females. They don't use this snout-up posture, but during roaring or bellowing displays they lift both their heads and tails out of the water, probably to let others see and appreciate their size. It could be another honest signal, like infrasound but addressed to those watching from the shore.

I was excited to find that gharials' "language" is very different from all other species'. Gharials produce long, monotonous buzzing, like small electric fans. Males also make amazingly loud "pong" sounds, apparently by slapping their jaws together just below the water surface. I was never able to see how exactly they do it. Their jaws are so narrow that slapping them shouldn't produce much sound. But gharials have two huge bony "bubbles" in their skulls, and these spheres probably work as resonators, making the sound so loud. Interestingly, the only other animals with such bony structures are kangaroo rats; the spheres apparently enable the rats to *hear* infrasound, such as the quiet vibrations produced by walking foxes and other desert predators.

It was clear to me that these "pong" sounds served the same function as headslapping displays of other crocodilians: they could spread for miles through the water, telling others where the animal making them was and probably also something about its size and strength. As could be expected for a species living only in large rivers, gharials are better at underwater signaling than at signaling through the air. "Pongs" carry much farther than "buzzes."

Gharials became a separate lineage around eighty million years ago, and their "language" is highly distinct. It occurred to me that you could probably figure out how closely related two species of crocodilians are by comparing their signals.

Comparing various human languages to find out what their "evolutionary tree" looks like is called *comparative linguistics*. Now I could start working on the comparative linguistics of crocodilians—something no one had ever tried before.

Crocodiles and alligators can understand each other very well. With a few exceptions, their postures, hisses, growls, and slaps have the same meaning. Their lineages may have split about seventy million years ago (just five million years before the extinction of dinosaurs), but their "languages" are still very similar. Human languages usually become mutually unintelligible just five hundred to seven hundred years after separation. What was so good about the crocodilian communication system that it could survive almost unchanged for so long?

Studying muggers and gharials in Katarniaghat was like working in a real-life *Jurassic Park*. I was surrounded with large, powerful ancient monsters that looked and sounded more or less the same as they had at the time of dinosaurs. One evening, as I was watching them from a tree, a huge Indian rhinoceros walked underneath, one of only two rhinos left in the reserve after decades of poaching. It was, of course, a mammal, and a relatively young species; the first rhinos appeared on Earth only forty million years ago. But it also looked like something from the Age of Reptiles.

When the mating season of both muggers and gharials was almost over, I packed my tent, walked to the railway station, and immediately found myself back in our time, the time of one species completely overrunning the planet. Nowhere is it as obvious as in India.

9
The Numbers Game
Alligator mississippiensis

Smoking is one of the leading causes of all statistics.
—Liza Minnelli

It was early April. Two weeks until the beginning of alligator mating season in the Everglades, and two months until its end at the northern edge of the species' range, where summer arrives later. I had to plan my research very carefully to make the best of this time. I managed to get free from teaching by making an arrangement with a friend: I taught double the normal load for the first half of the semester, and he took over for the second half.

I already knew an area where all alligators lived in small forest ponds: Fakahatchee Strand, the largest patch of tropical forest in the continental United States. Although huge and wild by Florida standards, it can be walked across in a day. There are a few canals along its borders, but alligators don't use these canals much, probably because of sharp limestone blocks lining their edges.

Now I had to find a place where all alligators lived in large lakes, rivers, or canals, preferably as far north as possible, so that the late onset of summer would give me some extra time. American alligators occur almost up to Virginia, but near the northern limit of their range they are generally rare, and getting a large sample size would be difficult. I ran a scouting trip along the Atlantic coast and found a great location: Savannah River Delta, a wetland bisected by the Georgia–South Carolina state line.

Before I could start regular observations, I had to answer one important question. My theory depended on the assumption that alligators could locate the source of underwater sound. This ability would make it possible for them to use headslaps as underwater location beacons. It's not a trivial thing to do. Most animals determine which direction the sound is coming from by using one or both of two methods. They can use the difference in loudness, created by the shading effect of the head on the ear located on the opposite side from the source of the sound. They can also use the minute difference in the time of arrival of the sound to the left and the right ear. Neither method works well in the water. Since living tissues have similar density to water, sound can easily enter your head and come out of it, so there is no shading effect. Also, the speed of sound in the water is 4.5 times higher than in the air, so the difference in the time of arrival of sound waves to your left and right ears is 4.5 times smaller. That's why people have difficulty telling which way a sound is coming from when they are diving.

Alligators are normally attracted to loud slaps on the water surface. It doesn't mean that they are always on the lookout for other gators making headslaps: splashes might also mean potential prey. I could use that habit to see if they'd be attracted to underwater slaps. Obviously, for a clean experiment I had to completely eliminate any above-the-water sound. But how can you make a slap on the water surface underwater? It was almost like the famous Buddhist *kōan:* "What is the sound of one hand clapping?"

At first I tried to make slap-like sounds by pushing inflated air balloons underwater and bursting them with a needle. Sounded very similar to a slap to me—but alligators didn't react at all.

So I decided to use a diving bell. It would allow me to create a small water surface that would be completely submerged. I assembled a contraption I named "the slapper." I took a heavy bucket, made it even heavier by wrapping a diving belt with thirty pounds of lead around it, and attached a thick metal pole to its bottom. Then I made a steel hook with a long handle and attached a plastic pad to the hook tip. Now all I had to do was push the bucket bottom-up underwater, stick the hook into the air bubble trapped inside the bucket, and make a slap with the pad.

It was very easy, and my first set of experiments was child's play. I'd go with two friends to a canal, find a group of alligators, and ask one of my friends to watch their movements. Then my other friend and I would take positions one hundred meters apart up and down the canal and walk into the water. One of us would be holding the slapper, the other a simple metal bucket. We'd both push our tools underwater, and the one with the slapper would make a slap. If at least one of the gators started swimming in the direction of the slap, we recorded the result as positive. There were a few rules to follow, like making sure the slapper wasn't always upstream, or always south, or always in the hands of the same person. We had to make sure the gators reacted to the slap and not to something else. Still, it took us just two weekends to run a few dozen tests and get a P-value of less than 0.001.

Later one of my colleagues pointed out that it would be better if the alligators had a choice of four directions rather than two. This experiment was much more difficult, because it had to be done in big lakes rather than narrow canals, and because alligators are seldom seen far from shore. Also, it required four people and four kayaks, three of which I had to rent. I eventually found a shallow area in Lake Okeechobee where it could be done, and over the next three years had to make a lot of weekend trips halfway across Florida with my friends and premed students (who'd let alligators bite their hands off for a good grade). By the time someone accidentally dropped the slapper into the lake's muddy depths, the P-value was below 0.004.

As soon as alligators in the Everglades started bellowing, I was off to Fakahatchee. The easiest way to get a large sample size would be to find the largest group of gators and count all their displays. But it would be completely wrong. It would constitute *pseudoreplication*, one of the most common mistakes in scientific research.

Let's say you want to know if men talk about sex more often than women. You watch one man for ten days, counting every time he mentions sex. Then you do the same with a woman. You run a statistical test with the data you have collected for ten man-days and ten woman-days, and get a good P-value. You have just committed pseudoreplication because your sample size wasn't 10 + 10. It was 1 + 1. You have data for only one man and one woman, and your results are virtually meaningless.

For me, avoiding pseudoreplication was extremely difficult because bellowing is contagious: if one alligator starts bellowing, others usually join him. I didn't want to count one bellowing chorus as many bellows. So I had to impose a lot of rules on the way I recorded my data.

In each pond, I had to choose just one alligator, and ignore all others. This would be my *focal animal*. I had to see it display once to make sure it produced infrasound, which meant that it was a male. I decided to limit my study to males because otherwise things would get too complicated. Then I'd record only bellows and headslaps made by this one male, and only if no other bellows or headslaps were made by him or any other gator within earshot in the previous hour. I also had to make sure that my focal animals were at least a kilometer apart, so they wouldn't influence each other. There were a few other restrictions, such as minimum size of the gator, maximum size of the pond, and so on, but I'll spare you the "Methods" chapter of my dissertation.

In practice it meant that I could watch only one alligator at a time and record not more than three displays in one sunny day. On rainy days alligators didn't do anything at all. My goal was to find ten qualifying males and record five displays from each of them. I'd watch them from sunrise till late morning, and spend the rest of the day looking for potential focal animals in other ponds. I already knew that it was better to look for small groups, always choose the largest animal first—it was most likely a male—and watch it only until the moment it left the water to bask in the sun, usually around 11 a.m. At night I watched alligator "dances" in larger ponds.

I spent four weeks in the forest. It was three hours' drive from my home, so I could see Kami only once or twice a week. She still insisted that we had broken up and wouldn't spend more than a few hours with me. In early May the rainy season began, but for now the rains were limited to the afternoon hours, so I didn't care. I got my fifty observations and drove to Savannah River.

This was an easier place to work. There was little forest, and I could drive on levees or paddle instead of walking for miles through dense jungle. I could even drive to Savannah, a big city across the river, if I wanted civilization—but I didn't. If I had a few free hours, I explored coastal marshes, where remnants of colorful Gullah culture still survived in small towns. The Gullahs (prior to that trip I had never heard of them) preserve more African customs and beliefs than any other African American group (excluding recent immigrants, of course). I learned, for example, that they consider large skink lizards deadly poisonous, just as many tribes of Nigeria and Cameroon do. Like people all over Africa, they tell funny tales about the smart and cunning rabbit fooling his predators.

Another interesting place nearby was Savannah River Site, a huge nature reserve surrounding a nuclear enrichment facility. Technically, almost all of it is closed to the public: you aren't even allowed to stop if you drive the only public highway crossing it. But this area has so much wildlife that simply driving through late at night can be a great experience. Once I even saw a Carolina dog, the rarest wild animal of the eastern States. It's a descendant of dingo-like dogs brought over from Siberia by the first Native Americans, but it has lived on its own in deep woods for centuries, if not millennia.

Camping wasn't allowed in the delta, and I didn't want to waste time driving to campgrounds, so I had to sleep in my car. In May nights were still cool, but in June it was too hot to sleep with the windows closed, so I had to wake up twice a night to put on a new layer of mosquito repellent. By the time I got my fifty displays and could return to Florida, I was completely exhausted.

I drove straight home, took a shower, and called Kami. And as soon as she spoke the first few words, I knew from the tone of her voice that she had found someone else. It was over.

I stayed in bed for two days, feeding on canned food from my hurricane stash. Then I went to see Steve. We looked at the results of the statistical tests. They were not what I expected.

In southern Florida, alligators produced almost no headslaps. In South Carolina, headslaps made up almost a half of all displays. This matched my predictions, and the P-value was good enough. But I also analyzed my data in a different way, counting the number of bellows each alligator produced in the first two days of observations, and found that there was no difference in bellowing between the two places.

Steve also pointed out another problem. How could I be sure that the observed difference in behavior was caused by difference in habitat? There were so many alternative explanations! Animals sometimes react to things you'd never think about. It could be climate, length of daylight, population density—many things. Mistaking correlation for causation is another error commonly made by scientists. You hear things like "Kids who often play violent video games commit more crimes," but what if these kids were just more prone to violence to begin with, or had issues with their parents, or lived under power lines, or didn't eat their vegetables?

"How can I eliminate all alternative explanations?" I asked.

"That's the most difficult part," said Steve. "You must at least eliminate as many as possible. You have four months before defending your thesis proposal. That's enough time to think of something."

I figured out how to do it in just a week, but the method I came up with wasn't an easy one. I had to find more places with just one type of habitat—either small ponds or large rivers and lakes. And they had to be located across the alligator's range in such a way that if you divided it into four quadrants, northeastern and so on, you'd have one place with each type of habitat in each of the quadrants. That setup would let me eliminate climate and daylight as alternative explanations. I'd look only at alligators found alone or in small groups, and avoid places with particularly high concentrations of gators. That would eliminate population density.

It was a beautiful setup, but it meant that I had to get data from six more sites. Two sites had taken me two months—the whole length of the alligators' mating season. Three more years . . .

And I had no idea if those six sites even existed. Finding an area where alligators live only in small ponds or only in large lakes and rivers is not as easy as it sounds. Being habitat generalists, gators would happily live in almost any body of water. Try finding a place with lots of small ponds but no rivers, or with large lakes but no small ones.

The rule was that once you defended your thesis proposal, you didn't really have to stick to it. You could modify it all you wanted, or even abandon it, change the subject completely and start from scratch—assuming, of course, that you didn't mind spending ten years in graduate school. So if everything went wrong, I'd have a second chance. That was good to know.

I decided to talk about it with John Thorbjarnarson. He knew as much as anyone about crocodiles and was always ready to help. I drove to Gainesville, where he lived, and showed him my results and my plans.

"That's too much work," he said. "Can you somehow move the animals from one habitat to another and see if they would change their displays? I know a few people who do crocodile translocations."

I looked into this possibility and found from literature that moving wild crocodilians from one place to another was very stressful for them. I'd have to wait for a year before I could be more or less sure that their behavior was back to normal. But to do that, I'd have to somehow mark them so that I could find them again after a year. To mark wild animals, you need special permits, which would also take a lot of time to obtain.

I was sure there had to be another way, but I didn't have time to follow this line of thought any further. I had to get on a plane and fly to a place totally different from Southern swamps . . . except it also had plenty of mosquitoes. It was a short trip that had taken two years to organize.

10
Crocodiles in Permafrost
Tagarosuchus kulemzini

When you love South, it's for its climate, its color, its diversity.
With North, it's pure romance.
—Valery Yankovsky

TWO YEARS EARLIER, the Danish Ornithological Society had asked me to help organize an expedition to Chukotka, the northeastern tip of Siberia. It's a vast, sparsely populated province, mostly covered with tundra and low mountains, not unlike northern Alaska, which sits across the Bering Sea.

This remotest part of the country has a colorful history even by Russian standards. The warlike Chukchi people invaded it many centuries ago and all but killed off the other tribes. There are still a few Yupik Eskimo villages in the east and a handful of Kereks, Yukaghirs, Lamuts, Koryaks, and Evens in the far south and west.

The first Russians to map this land were Siberian Cossacks, an ethnic group remarkably similar in lifestyle and function to the early Texas Rangers of the Comanche Wars. Despite centuries of conflict, the Russians were unable to fully subjugate the Chukchi, who remained the only tribe free from paying taxes to the czar in the Russian empire. In the late nineteenth century Chukotka had stronger trade ties with the US than with Russia; some old people in coastal villages still speak a bit of English and quietly lament the sad fact that this land wasn't sold to the US together with Alaska.

In the Soviet era, explorers found ore deposits (now most mines are closed due to high operation costs). All shamans were executed, and children were taken to boarding schools, but, unlike many other ethnic groups of Siberia, the Chukchi preserved much of their splendid traditional culture based on reindeer herding and marine mammal hunting. In some villages they've managed to resist the lure of alcohol. In many others they haven't, and the consequences are as bad as it gets. They are supervised by disproportionately abundant government bureaucrats and border guard troops.

A few thousand Russians have hung on in district centers since the times of Soviet colonization of the North, mostly as drivers, seamen, gold miners, and doctors. A few of them fell in love with Chukotka and desperately try to protect it, but others treat the land as an occupied territory, shooting and trapping anything that moves. There are also two small and remote Cossack villages.

Chukotka is largely neglected by Moscow but is not exempt from a constant stream of stupid laws and regulations aimed at facilitating the thorough corruption on which the Russian government depends for survival.

My guests were the first Western bird-watchers to visit Chukotka in ten years. In the mid-1990s, tourists from another European country chartered a helicopter there. Halfway along the route, the pilot landed on the tundra and threatened to leave them there unless they paid him $3,000 per person. I don't know how much they ended up paying, but their story killed all ecotourism in the region for a long time.

Our main goal was to find a tiny bird called the spoon-billed sandpiper. It breeds only on the remotest shores of Siberia and is now on the verge of extinction because of traditional hunting with mist nets in Myanmar, where it winters. At the time of our visit there were about two hundred birds left; now there are far fewer.

Organizing this trip took two years. We had support from official and unofficial authorities and were very lucky with weather and transportation (both are difficult-to-win lotteries in Chukotka). Still, nothing went as planned. Even in the most corrupt parts of Africa you'd have to search hard to find such all-encompassing inefficiency. Feudal Soviet bureaucracy is covered by a thin gloss of "wild" capitalism. To visit a relative in a village fifty miles away, some local fishermen have to fill ten pages of paperwork and pay $3,000 for helicopter tickets. They can't use their boats because independent travel to another district is not allowed in the "special border zone," which stretches for hundreds of miles inland from the entire Arctic coast. And even if they buy those tickets, they might have to wait for months for a flight, only to get bumped off it at the last moment because some official's brother-in-law wishes to fly to a remote island to hunt endangered walruses with a machine gun.

My young Russian colleague Ivan Taldenkov and I had to obtain twelve different permits. It wasn't enough. Local officials would hunt us down every time we had to pass through a town, and demand that we purchase more permits. At some point Ivan lost his cool and yelled at them: "Leave us alone! We got twenty permits already! We can go anywhere we want! This is a free country!" The officials fell silent for a second, then one of them said in disbelief: "Who told you such a stupid thing?"

Nevertheless, the expedition was a success. In just three weeks in July, we found almost every bird species known to breed in Chukotka, plus a few birds never before seen there. Most of those were recent arrivals, brought by global warming. Huge stretches of lichen tundra are now being overgrown with tall shrubs, and the forests are marching north at five miles per year.

The Arctic, slowly recovering after decades of Soviet abuse, was even more beautiful than we expected. For hundreds upon hundreds of miles, the tundra looked like an immense flower bed, with crazy washes of scarlet, purple, blue, and yellow running down the hills all the way to the coast, where immaculate white blocks of ice were slowly drifting by, as if soaring between sky-blue sea and stormy-blue sky. At midnight the sun was so low in the sky that the ground was almost in shade, but the flowers were still brightly lit, like little lanterns.

My guests were hard-core birders. Most of them were well traveled and prepared to go through almost anything to find a rare bird. But even they were totally overwhelmed by the complete unpredictability of our journey through that unearthly wilderness. It was just too much for a logically thinking person: the squalor of semi-abandoned towns lost in the cosmic expanse of wild bays and floodplains, rotting boxes of radioactive fuel left decades ago near defunct automatic lighthouses, rusty gasoline canisters scattered everywhere from tiny islands to snow-clad mountaintops, middens of recently extinct native villages surrounded by the solemn grandeur of deep fjords.

For me it was a wonderful but sad experience. Not only because I saw my native country being sucked dry by a handful of thieves, but also because we did eventually find a nest of spoon-billed sandpipers. We visited many coastal spits where they had lived just a few years earlier, but all were empty, except one. I held three tiny, beautifully spotted hatchlings in my hand for a few seconds while Ivan was preparing to band them. They were almost weightless, fluffy and fragile like dandelion heads, with bright black eyes, soft spoonlike bills, and warm bellies. They were peeping quietly. It was hard to believe that these gentle babies were capable of surviving ferocious Arctic storms, summer blizzards, marauding predators, and the hardships of the journey they were to embark on in just four weeks, ten thousand miles round-trip. And these brave little wonders were going to be extinct in just a few years.

I was hoping that during this month-long break from fieldwork I'd have time to think about scientific problems. But I was too busy navigating our way through bureaucratic barriers, crisscrossing the tundra in search of rare birds, and making sure our supplies of smoked salmon and reindeer venison didn't get left behind during countless transfers from boats to choppers to caterpillar ATVs. Only when we returned to Anadyr, the provincial capital, did I finally have a chance to get back to crocodiles.

Anadyr sits on a hill overlooking a large bay. The airport is on the other side of the bay, a few miles away. To make the crossing, you have to take a barge (in the summer), a bus shuttling on an ice road (in the winter), or a helicopter (in the fall and spring). The latter option might cost more than the flight from Moscow. The town is made up of a sorry bunch of Soviet-era apartment blocks, many of them abandoned but all painted gaudy colors to celebrate the 2007 visit by Vladimir Putin, the country's "godfather." On Tuesdays and Sundays there's no transportation at all: Tuesday crossings were banned by a superstitious city mayor after one rusty barge sank on a Tuesday with all aboard, and Sundays are out because of the governmental policy of forcing Orthodox Christianity down everybody's throat. So if your plane lands on Tuesday or Sunday, you either have to pay $200 for a hotel room or seek shelter in one of the abandoned buildings. However, getting stuck at the airport in the summer has its perks: there's a colony of rare Aleutian terns just outside the terminal and a huge gathering of beluga whales around the barge pier.

Once you make it to Anadyr, there are two places worth seeing: the gargantuan state-built Orthodox cathedral, assembled in 2003 with timber transported seven thousand miles, and the provincial museum with a stunning collection of local art—mostly carved walrus tusks, mammoth ivory, and whale vertebrae. There's also a small paleontological exhibit in the museum, with plenty of Ice Age stuff and one modest dinosaur bone.

Throughout the Age of Reptiles, Siberia was warm enough for dinosaurs and sometimes for crocodiles. The best known Siberian crocodilian fossil is *Tagarosuchus*—a small, agile, long-legged terrestrial predator somewhat reminiscent of a greyhound.

The diversity of crocodilians at that time matched the diversity of their relatives the dinosaurs. Some were entirely marine, with fishlike tails and paddle-like feet, but many were terrestrial. A few evolved into herbivores, and some could climb trees. The smallest forms were just a foot or two long; the largest could grow to at least forty feet. *Saurosuchus*, a humongous crocodile from what is now the Sahara, fed on fish while young but switched to hunting dinosaurs when it grew large enough. It weighed up to eight tons and could easily drag almost any dinosaur underwater. *Deinosuchus*, a colossal alligator of similar size, hunted dinosaurs in eastern North America.

When the mass extinction came, something really strange happened. Dinosaurs were completely wiped out, except for one group known today as birds. But a few crocodilian lineages survived. Most, if not all of them, were aquatic. It's possible that one terrestrial group had also survived: terrestrial crocodiles occurred in Australia and neighboring islands until they were hunted out by people just a few thousand years ago. It's unknown if they were mass-extinction survivors or a more recent split from aquatic crocodiles.

Why were crocodiles able to survive while non-bird dinosaurs were not? No proposed theory of the causes of the mass extinction (there are many) explains it well. It probably had something to do with the amazing fasting ability of crocodiles. Large individuals can last for up to a year without food and generally eat about ten times less than a mammal of the same size. However, small birds and mammals almost certainly had to eat much more than dinosaurs (per pound, of course), and they also survived, so there's no straightforward explanation.

At least it's easy to explain why the largest crocodiles didn't make it. There were no dinosaurs left for them to hunt, and large mammals didn't appear until millions of years after the mass extinction. But what happened to the smallest crocs? Crocodilians today are unique among animals in having no species smaller than four feet long. There are rabbit-sized hoofed animals, large-watermelon-sized cetaceans, and minnow-sized sharks. But there are no really small crocodiles.

There's no physiological or ecological reason. Hatchlings of dwarf caimans are less than ten inches long and survive well in predator-filled Amazonian swamps. As Herodotus wrote in his description of the Nile crocodile, "No other animal begins its life so tiny and grows to be so huge." But in all living crocodilians, females are at least three feet long at the time of puberty, and males at least four feet.

A ten-hour flight from Anadyr to Moscow was all it took to find a place with a decent Internet connection. I had come up with an idea, and I was dying to discuss it with experts. Darren Naish, a famous British paleontologist, told me that although there were plenty of very small crocodilians during the Age of Reptiles, none of them belonged to the so-called "crown group"—the branch of the crocodilian evolutionary tree that includes all surviving species.

Today's male crocodiles use infrasound as an honest signal to impress skeptical females. Producing infrasound requires immense physical strength, and a crocodile has to reach a certain size to be able to do it. What if this requirement prevented crocodilians from becoming smaller? What if the "crown group" crocodilians evolved using infrasound and then couldn't stop doing so, because their females would ignore infrasound-challenged males?

If true, this would be the first known case of female choosiness limiting the evolution of an entire group of animals. But it would be very difficult to prove this theory. Nothing is known about the signals used by extinct species. The giant dinosaur-hunting *Saurosuchus* from the Sahara had a bizarre growth on its snout, possibly used in sound production—but even that is uncertain.

To check if that theory was at least plausible, I had to find out if the smallest of today's crocodilians could produce infrasound. Alligators become capable of doing so when they are about four feet long. Males of the smallest crocodiles and caimans reach maturity at approximately the same size. But very little was known about them.

I had a lot of work to do.

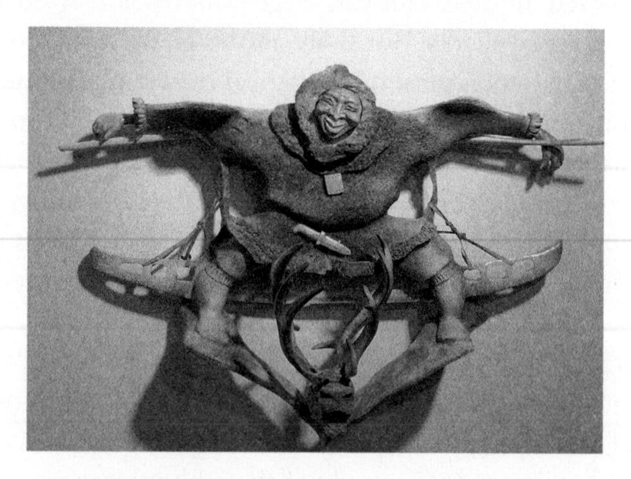

Whale vertebra carving by a Chukchi artist

11
River World
Caiman niger

The happiest moments of a jungle expedition are when you enter the jungle and when you exit it.
—Arkady Fiedler

My adventures that summer were just beginning. The crocodilians that were supposed to have their mating season in July and August were saltwater crocodiles in Australia, Philippine crocodiles, and three caiman species in the central and eastern Amazon. With the money I made guiding the Chukotka expedition, I could go only for the caimans.

The cheapest place to fly to in South America was Caracas. From there I took a bus to Santa Elena de Uairén, a small town in the far southeast of Venezuela. The bus ride took sixteen hours; I carelessly put my backpack into the luggage compartment and almost froze to death. Gasoline is sold for a few cents per gallon in Venezuela, so bus drivers aren't concerned with fuel economy and air-condition their buses almost to the freezing point. They believe that arctic cold will help them stay awake.

Santa Elena is on the Brazilian border, so, as you might have already guessed, its most striking feature is long queues of Brazilian drivers visiting Venezuela to fill their tanks. Their cars have mostly been converted into huge tanks on wheels.

The town is the starting point for climbing Mount Roraima. It's one of the world's most spectacular hikes, so there are lots of tourists there. But, although Santa Elena can be reached from Miami in two days, almost all tourists are South Americans, Europeans, or Israelis. Americans are strangely underrepresented almost everywhere in the world, except for a few nasty tourist traps like Cancún, the Bahamas, and Cabo San Lucas. I guess it has something to do with short vacations and the abundance of scary TV programs with titles like *Locked Up Abroad*.

For me, Santa Elena was a convenient access point for the Brazilian Amazon. A good highway cuts from here all the way south to Manaus, the unofficial capital of the immense Amazon Basin. There the road ends. Maps show highways fanning out from Manaus in all directions—east to the delta, south to more civilized parts of Brazil, and west to Peruvian, Bolivian, and Colombian borders. But these roads either don't exist or are barely passable by tractors and ATVs during the dry season. The government stopped almost all road construction in the Amazon when it became clear that every new road causes immediate destruction of all surrounding forest by illegal loggers and desperate, landless farmers.

So virtually all travel in the Amazon is by boat. It's a well-developed and efficient system. People just bring their hammocks and hang them in long rows on the deck. It takes up to two weeks to get to the more remote river ports. The journey can get boring despite colorful people around you, gorgeous sunsets, and pleasant weather. You don't see much wildlife, except for two species of dolphins and a few birds. The river is too wide, and most large animals near its shores have been hunted out. It's so frustrating to watch murky water and green jungle for days, knowing that countless wonders are hidden in the river and the forest, but never seeing them.

I traveled upstream, trying to learn as much Portuguese as I could. I had an informal agreement with the administration of Mamirauá, a large nature reserve at the confluence of Rio Solimões (as the part of the Amazon River above Manaus is called in Brazil) and Rio Japurá. They promised to let me study black caimans there.

I arrived in Tefé, a city near the reserve, on a Friday evening and had to wander its hot streets for two days, waiting for the reserve office to open. When it opened, I was told that I'd have to pay "the tourist price" to stay in Mamirauá, $500 a day. All the money I made guiding the month-long trip to Chukotka would cover just two days.

Of course, I couldn't give up. I needed data on black caimans, and there are only a few places in South America where they are common and easy to observe.

Just two centuries ago, the black caiman was the most common of five caiman species in the Amazon. But it also had the most valuable skin of them all. Crocodilians have small bones called osteoderms embedded in their belly armor, and these bones make the skins difficult to work with. In different species osteoderms develop to various extents, but black caimans have the least of all caimans. So they were sought out by hunters and soon became vanishingly rare throughout their extensive range. Not even the remotest rain forest is completely safe for animals nowadays.

More recently, all countries that had black caimans declared them a protected species and banned skin exports. Crocodilians are extremely resilient and, given a chance, can rapidly bounce back from the brink of extinction. Unlike some other species, which are captive-bred in hundreds but have no good natural habitat left, black caimans still have plenty of living space. They are now becoming common again in remote parts of Brazil and neighboring countries. Brazil and Bolivia are even developing sustainable hunting programs.

I took a boat to a small village across the river from the reserve and rented a dugout canoe from an old *caboclo* (riverside resident) who lived in a stilt hut off the beach. Now I could easily cross into the reserve and observe the caimans, but I didn't want to go illegally. I always think that as a biologist, I must follow all conservation laws and rules meticulously—otherwise I have no moral right to expect the same from others. So every evening I'd cross the river and patrol the very edge of the reserve, looking for caimans from the outside.

Most of Mamirauá is flooded to the depth of ten feet when the Amazon is at its highest level in June. Now it was already down a bit, but thousands upon thousands of square miles of riverside forest were still underwater. I never miss a chance to snorkel in tropical rivers and lakes if the water is not completely opaque. Snorkeling in Mamirauá was exhilarating. Unlike in the river itself, the water between the trees was clear, so I could see pink dolphins, huge turtles, and small black Amazonian manatees. Fish were fascinating: tiny fluorescent tetras, beautiful discuses, lethal electric eels the size of a log, splendid blue stingrays that looked like big butterflies. Giant armored catfish were leaving deep trenches in the muddy bottom. They looked scary with their mean faces, spiny plates, and huge mouths but were completely harmless. Once I witnessed the majestic mating dance of two eight-foot-long arapaimas, the largest fish in the Amazon. There was plenty of wildlife above the water, too. I even saw the rarest and weirdest of all South American monkeys (and there are some really weird monkeys in South America), the white uakari with its long white hair and bright-red bald head and face.

But I saw hardly any caimans. Almost all of them were apparently hiding deep in the forest, away from big rivers with their boat traffic, noise, and poachers. Finally, I found a female with a bunch of tiny, angelic-looking hatchlings. This was a bad sign. The mating season was almost certainly over. I had to go east, toward the Atlantic coast, where the rainy season ends a month or two later.

I returned to Tefé and got on a boat to Macapá, the last big city on the northern side of the Amazon before its exit from the forest into the Atlantic. It took a full week. Summer was slipping away fast.

The only interesting thing I saw that week was *Encontro das Águas* ("Meeting of Waters") just below Manaus, where the black waters of Rio Negro join the light-colored waters of the Amazon and flow side by side for almost a hundred miles without mixing. Rivers of the Amazon come in three colors: pale-tan whitewater rivers flowing from the Andes and rich in sediment; tea-colored blackwater rivers that emerge from dense forests and get their color from rotting leaves; and rare, precious clearwater streams that come from sandy or rocky areas. Finding a nice clearwater river is a real treat. Usually, they are the only ones where you can snorkel. Wildlife is obviously different in these three kinds of rivers, and subtly different in the forests on their shores.

I didn't know what to do next. There were black caimans in French Guiana and Suriname, but it would be difficult to arrange cheap transport to their remote habitats since I didn't speak much French or any Dutch. So I crossed both countries and got to English-speaking Guyana, a small country in northeastern South America. Here I found Peter Taylor, one of the world's leading experts on black caimans, who just happened to be in Georgetown (the capital) on a break from his work in the interior. Peter is known for his unconventional approach to field research: he has managed to engage an entire Indian village, training its residents as naturalists and tour guides for Rupununi Learners, his caiman study center.

Peter said that my timing was perfect. Three of Guyana's four caiman species were mating in August. He recommended flying to Karanambu Ranch, where a large population of black caimans still existed. Finally, after almost a month in South America, I was getting somewhere. But time was running out. I had to be back in Miami in about a week. And getting there was going to take a while. I had a return ticket from Caracas, but couldn't go straight to Venezuela, because there were no roads and no flights. Venezuela claims more than a half of Guyana as its own territory, and the relations between the two countries are unfriendly, to put it mildly.

Thirteen years earlier, when I first went to South America, almost every country there had territorial disputes with some of its neighbors. Bloody wars over land were frequent in the history of the continent. At one point in the nineteenth century, almost all the male population of Paraguay fell in battles. Each country had its own version of maps and would meticulously mention all claimed territories as its own in weather forecasts. But in the late 1990s most of these disputes were settled one way or another. Only two bastions of stupid stubbornness remain: Argentina's completely unjustified claim to the Falkland Islands and Venezuela's ridiculous claim to most of Guyana.

So my choices were to go through Brazil or fly to Caracas via Trinidad. Either way, I had just three days left in Guyana.

And that's when I got my stay of execution. It was an email from my department, telling me that I was awarded a Tropical Biology fellowship. They granted only one each year—and I won it just when I most needed it. It meant that I would get my regular stipend for one year without having to teach. If I was thrifty in the extreme, I could even stretch that money for two years.

It also meant that some very experienced biologists considered my project worth supporting. I wasn't completely sure about it myself at that point. So their trust in me was really encouraging.

Of course, I still had to return to Miami in about a month to write and defend my dissertation proposal, but a month seemed like a lot of time. I called Karanambu Ranch, the place recommended by Peter, and hopped on a plane to Lethem on the Brazilian border. From there another plane took me to a landing strip in Rupununi Savannah. To my surprise, both times I managed to talk the pilots into letting me fly the planes, even though the first one had twenty-something passengers. I had never had a chance to fly and land anything larger than a Cessna before.

This area is a transition zone between the great Amazon rain forest and the savannas of the Guiana Highlands. From the air it looks like a tangle of giant snakes: widely meandering rivers are lined with broad belts of riparian forest. It was the rainy season, and much of the country was flooded. I managed a nice landing, the plane left, and I found myself alone in the sea of wet grass. There was only one trail leading away from the landing strip, so I picked up my backpack and followed it toward the distant forest. As soon as I got under the first trees, the trail turned, and I saw a narrow channel—obviously a jeep road in drier times of year—a dugout canoe with a motor, and a woman sitting on its side. She was Diane McTurk, a living legend.

If you were a kid growing up in the Soviet Union in the 1970s and interested in nature, you invariably had Gerald Durrell as one of your most loved authors. His books are wonderfully written and full of great British humor, but, strangely, they are less known in the US than in most other civilized countries. His best books were written from 1950 to 1980, so they were published in Russia and became immensely popular. Books written after 1980 were almost never translated to Russian, because in that year Russia joined the Universal Copyright Convention and books by foreign authors couldn't be published anymore without paying royalties. I only got a chance to read Durrell's last works after moving to the US. But his book that I happened to read first, at the age of about five, was about Guyana. It was called *Three Tickets to Adventure*, and Diane McTurk was one of its most memorable characters, a daredevil woman with long black hair who rode wild buffaloes and wrestled giant anteaters with her bare hands.

Diane is a daughter of a pioneer cattle rancher who founded Karanambu in 1927. She's now in her eighties, but she runs the ranch (which has mostly been turned into a private nature reserve), a conservation trust, and a rehabilitation center for giant otters. She also guides wildlife tours through surrounding forests by motorboat, by jeep, or on horseback. Just a week before my arrival, she was driving a jeep along the awful eighty-mile road from Lethem when a bridge collapsed and the jeep fell into the river. She was badly bruised and broke a hip, but that didn't stop her from showing me around, and "around" meant everything within twenty miles. She knows every bird, bat, or monkey within that area almost personally. Exploring the jungle in her company was a wonderful experience.

For the first time that summer, I could relax and do my job. No more bargaining with stupid bureaucrats, no more waiting for days to catch a boat or a helicopter, no more endless bus rides. And I was in one of the most beautiful places in the entire Amazon.

The flood made everything easily accessible. Oxbow lakes, usually hidden deep in the forest, could be reached by paddling along narrow trails. Impenetrable swamps had turned into crystalline pools. And every lake had at least one big black caiman, sometimes over twelve feet long. Juveniles were hiding in flooded forests and were much more difficult to find, but sometimes I saw their eyes reflecting my flashlight at night.

The oxbow lakes were sublime. They were covered with floating leaf pads of *Victoria* water lilies up to ten feet across. The flowers of *Victoria*, larger than my head, would silently open every evening, petals spreading so fast that their movement was easy to see. They were white on the first night and pink on the second. I've traveled all over the Amazon but have never seen so many of them. The lakes were very quiet at sunset, never a mosquito buzzing, only giant arapaimas splashing sometimes. I'd sit there in my canoe, slowly adjusting to the pace of the world around me. Observing caimans takes a lot of patience.

Black caimans are probably the most beautiful crocodilians. They are boldly patterned and dramatically sculptured, with huge intelligent eyes. Although not as large as some crocodiles, they are the largest members of the alligator family, which includes alligators and caimans. Unlike alligators and all other caimans, black caimans are fiercely territorial. They are also the only caimans known to kill humans (a few every year), but most attacks are probably territorial defense, not hunting for food. One large male, longer than my canoe, didn't like the boat when he first saw it, probably because it was sitting very low in the water with my head barely visible above the side. The caiman raised its heavy head like a cobra, gave a low, menacing roar, and splashed the water with his huge tail. But eventually he calmed down and before dawn rewarded my good manners by performing a roaring display, with infrasound so powerful that I could feel my boat vibrating despite being more than a hundred yards from him. Later I learned that black caimans did head-slapping displays as well. And, just like alligators, they would sometimes crawl out of the water at night and wait for passing animals on trailsides.

If I could have, I would have stayed at Karanambu for a few months. I don't know what I enjoyed more: talking to Diane and her charming family at dinner, playing with giant otters, cruising the flooded forests in search of rare monkeys, riding across the savanna looking for giant anteaters, snorkeling with angelfish among spiky stems of water lilies, or simply floating in the middle of a lake at sunset and watching those water lilies open. But I had to leave. I got basic information about black caiman displays; that was all I could hope to learn about them at the moment. I had to move to other parts of Guyana, where three other caiman species were supposed to live.

12

Vibrating Toy

Paleosuchus palpebrosus

The smaller the caiman, the sharper the teeth.
—Guyanese proverb

I RETURNED TO GEORGETOWN, the most modest-looking capital city in South America. This town and much of the country feel more like the Caribbean: very laid-back, relaxed, largely immune to the rash of development that has recently reached even the remotest parts of the continent.

Of course, there is some development in Guyana. Flying over the jungle you see plenty of recent clear-cuts and huge barren patches of toxic tailings from gold mines. But the impressive life improvement that is so obvious in Brazil, Peru, or Ecuador is yet to be felt here. I saw a few lepers begging for food in Georgetown—a very rare sight anywhere outside India nowadays.

One place in Guyana I wanted to visit was Adventure, a small town north of the capital. Durrell's *Three Tickets to Adventure*, the first book about wildlife I ever read, was named after it. I found Adventure much less appealing than it apparently was in Durrell's time, but there were still some spectacled caimans in marshes and irrigation ditches there. This is the most widespread caiman species, common from the Amazon all the way to southern Mexico. Its name comes from a bony ridge on the forehead that looks a bit like the bridge of spectacles—not a good name, because some other caimans also have it.

In places where both black and spectacled caimans occur, they occupy different habitats. In Peru black caimans are mostly found in oxbow lakes, while spectacled caimans live in rivers. In Brazil, black caimans choose large rivers, and spectacled caimans are more common in smaller ones. In Guyana, black caimans are the masters of forest lakes and rivers, while spectacled caimans inhabit savanna rivers, coastal lakes, and mangrove lagoons. I don't know why this habitat separation is arranged differently in different places. What was important for me was that both species were habitat generalists, so, according to my theory, they were supposed to use both roars and headslaps. I had already found that black caimans did just that. As for spectacled caimans, their mating season was six months away.

There are two more species in Guyana—the so-called dwarf caimans. They don't grow to more than six feet. In fact, one of them is the smallest living crocodilian: even adult males are about five feet long. Their habitats, too, differ in complicated ways in different parts of South America, but the general pattern is that they are usually found in fast upland rivers, small forest streams, flooded forests, and forest swamps, with one species preferring whitewater river basins, and the other being more common in and near blackwater rivers. The important part for me was that both species lived in both rivers and small ponds.

Dwarf caimans aren't particularly rare, but it's very difficult to find them when you need them. The most reliable way is to snorkel down small streams and look under overhanging banks, tree roots, and flooded logs, where these shy, placid creatures hide most of the time. But sometimes they leave the water to hunt on land, and they can occasionally be found wandering late at night in the streets of Manaus.

Peter Taylor told me about a population of dwarf caimans in a nature reserve called Iwokrama Rain Forest, one of the few places in the interior accessible by road during the rainy season. I was happy to find a study site that could be reached in just twenty hours without paying too much, and hopped on a bus going in that direction. Just fifty miles out of town, we stopped at a police checkpoint. There was a long queue of trucks and buses ahead of us. Every passenger or driver had to slowly, painstakingly dictate his name, address, and occupation details to a very drunk police captain who had trouble holding a pen. All luggage (and everybody had a lot of it) had to be searched.

When my turn came after a couple hours of waiting, the drunk captain searched my backpack personally. He got to my small medical kit (traveling in such places, you are frequently asked to help even if you aren't a doctor, so I used to carry some basic stuff like Metronidazole and Coartem) and fished out a tiny vial of nitroglycerin pills.

"Aha!" he said. "This is explosive! You are terrorist, man! You stay here under arrest and go to Georgetown jail tomorrow!"

It was obvious that he was asking for a bribe in front of at least a hundred people. I couldn't afford a bribe, so I started arguing, waiting for him to look the other way so I could quietly roll the vial into one of many seemingly bottomless cracks in the floor. Everybody, including other policemen, was looking at the captain with intense hatred. The country's main highway remained paralyzed. Finally, a bespectacled guy wearing a necktie stepped out of the crowd and said:

"I'm a doctor. What's the problem?"

I explained. The guy smelled a pill from the vial, pretended to be analyzing the smell, and said, to everybody's amazement:

"You bloody idiot, comandante! These are heart pills! You are drunk!"

The captain turned red but didn't say a word. I threw my stuff back into the backpack and returned to the bus. Along the way I caught up with the nitroglycerin expert walking to his car.

"Are you really a doctor?" I asked.

"No way, man! Me, I'm real estate salesman."

Iwokrama Rain Forest turned out to be a nice reserve, with lots of swamp forests and little streams perfect for dwarf caimans. Spectacled caimans were easy to see in the main river, but dwarf caimans lived deep inside the forest, in shallow puddles left by the receding flood. These were Cuvier's dwarf caimans, the smaller of the two species and the smallest living crocodilians. During the day they hid in deep burrows under the roots of big trees. These "toy" caimans are good diggers, and their head is shaped for that: short, heavily armored, with the snout slightly upturned and the protruding upper teeth forming kind of a rake. They are very beautifully patterned, with chocolate spots, white bands, and big brown eyes. The babies are especially cute.

The largest dwarf caiman I found was five feet long; only male "dwarfies" grow that big. I spent a long night watching him while he happily fed on countless fish trapped in the puddle, and countless mosquitoes happily fed on me. It was a beautiful night. The entire forest floor was brightly lit with a cold, ghostly glow, white or pale green. Iwokrama has unusually abundant luminescent fungi.

The caiman "sang" only once. He raised his head and tail and made a sound resembling a dog's growl. I thought I saw his back vibrate (which would indicate he was making infrasound) but couldn't be sure.

I returned to Georgetown and—by a happy coincidence—found I had an email from Colin Stevenson, an Australian herpetologist who is one of very few people in the world you could call a dwarf caiman expert. Colin wrote that, according to observations made in zoos, dwarf caimans produced barks, headslaps, and infrasound. Of course, it didn't mean that my search for them in the forest was unnecessary. I already knew that crocodilian behavior in captivity and in the wild could be very different.

This new information matched the predictions of my theory. But it was even more interesting to learn that even these "toy" crocodilians could produce infrasound despite being so small.

A few days later I got another email. It was from Nick, the only American tourist I ever met in Guyana. We had spent a few hours together on a bus on our way from Iwokrama to Georgetown. Nick was a student from Indiana who came to Guyana to study folk medicine. He never managed to obtain all the permits required for visiting Indian lands, so he was limited to interviewing mestizo settlers who didn't know that much about herbs.

"You are a very brave man," he wrote. "You aren't afraid to walk in the forest with all those caimans. And when we arrived to Georgetown, I saw you give a coin to a leper. How could you touch his hand? I'd rather die."

I found it ironic. Not because the largest caiman I saw in Iwokrama was much smaller than me, and not because leprosy is not particularly contagious, is not transmitted through touch, and is easily treatable with antibiotics. In the next paragraph Nick wrote: "I got a tattoo in Georgetown, it covers all my back. I always wanted such a tattoo. It even has a caiman in it! I found a small place behind the docks where they did it really cheap!"

Call me a coward, but I'd never go to a "really cheap" tattoo parlor in Georgetown. Nick's courage—or whatever you'd call it—was really amazing.

I flew to Trinidad and spent a few days in this island paradise before returning to Caracas and then to Miami. Between hiking in mountain forests and watching baby sea turtles hatch on moonlit beaches, I did some paddling through mangrove lagoons in search of the pygmy anteater, an animal so sweet-looking that its local names in many parts of South America mean "little angel." I found only one anteater but plenty of spectacled caimans, and noticed that they were smaller than those on the mainland. A year later I learned that on the neighboring island of Tobago they are even smaller—adult males are only five to six feet long. I found a bunch of Tobago caimans in a zoo in South Carolina and in just one morning of observing learned that they, too, produced roars, headslaps, and infrasound.

Eventually I confirmed that sexually mature males of all small crocodilians can produce infrasound once they reach four feet in length. Apparently, four feet is the minimum size required to start giving that important honest signal of strength and maturity. Is that why no crocodiles today are smaller than that? Seems very likely, but I still can't think of any practical way to test this hypothesis.

13
Tales of Love and Friendship
Caiman yacare

Nothing's better than seeing a familiar face in a foreign land.
—Brazilian proverb

I SPENT SEPTEMBER IN MIAMI, passing qualifying exams and working on my thesis proposal: going over the information I had gathered, summarizing the results, and planning ahead.

I already had some data on seven species, and so far the predictions of my "master theory" seemed to be accurate. Species that were habitat generalists regularly used both vocal signals (bellows or roars, whatever you call them) and slaps. Species inhabiting only large bodies of water—rivers and mangrove lagoons—used mostly slaps. The Chinese alligator, which inhabits small lakes and ponds, used a lot of bellows but almost no slaps. I was going to try observing more species, of course.

My results from two populations of the American alligator were less clear. Animals living in small ponds used a lot fewer slaps than those living in rivers and canals. But their bellowing activity was exactly the same. Also, there were lots of possible alternative explanations for the difference in headslapping other than the habitat structure.

To clarify the situation, I was going to find more populations living either only in small ponds or only in big lakes and rivers. Ideally, their geographical locations would form a checkerboard pattern. That would eliminate many alternative explanations—the effects of climate, daylight hours, and so on.

I also decided to do a similar study on another species. I just wasn't comfortable having to extrapolate data from just one species to all of them. There was only one other crocodilian fitting the requirements: common, with a large range and broad habitat preferences. But this one would be a lot more challenging to study than the American alligator. It was the Nile crocodile.

In addition to testing my "master theory," I wanted to answer another question. It seemed that my idea was correct, and the composition of crocodilian "songs" depended on habitat. But how did it work? Did every male alligator wake up in the morning, look around, and decide, "I'm in a small pool, so I'll bellow a lot but headslap only once a week"? Or maybe he didn't change his behavior at all, but inherited that particular way of signaling from generations of his ancestors that also lived in small ponds? It was the old nature versus nurture problem.

The easiest way to find out would be to catch a few dozen alligators living in small "gator holes," move them to large lakes, and see if they'd begin to headslap more often. But that would be totally impractical for a number of reasons.

It occurred to me that instead of moving the animals between lakes, I could change the size of the lakes they were living in. I considered using artificial reservoirs, then realized that some lakes changed their size naturally, shrinking during droughts or expanding during floods. And then I remembered yacare caimans.

Yacare caimans live in South America, from the southernmost parts of the Amazon Basin down to northern Argentina. Most of their range is in tropical savanna, which is wet or flooded during the rainy season but dry for the rest of the year. Their mating season falls in the transition time from dry to rainy, so there should be abrupt changes in water levels. In Bolivia I noticed that yacare caimans have unique markings on their jaws, making it easy to recognize individual animals. So I could come back to the same lake after it changed size and get a second set of observations of the same individual caimans.

It took a month of searching through obscure hydrological journals and consulting with my friends all over South America, but by the end of September I had a beautiful plan for a six-week yacare study.

Yacares mate from mid-October until mid-December. In most of their range the rainy season starts during that time, so the water levels go up. But in the Pantanal, a huge seasonal wetland shared by Brazil and Bolivia, the flooding happens much later, because the water causing it comes from Mato Grosso Plateau farther north and needs time to reach the Pantanal once the rains begin.

So I could have two study sites, one in the Pantanal and one elsewhere. In the Pantanal the water levels would be slowly dropping. I'd find a few very shallow lakes, observe the caimans there for a week, then wait until the lakes would partially dry and break up into small ponds, and observe the same caimans again to see if they'd begin to use fewer headslaps and more roars. At the second study site, I'd do the same, but the water levels would be rising and small lakes would grow into big ones, so I should see opposite changes in caiman behavior.

Of course, there were lots of possible reasons for this plan to fail. But in field zoology you can never be sure of anything. Even the most perfectly designed study can go terribly wrong simply because the rains come too late or too early, or the animals get wiped out by a disease outbreak, or you get sick and miss two months of work, or your study subjects simply don't behave the way you expect. It happens all the time; people have wasted huge research grants and decades of their life simply because of bad luck.

I successfully defended my proposal, so all bets were on. I followed the familiar route: a cheap flight to Caracas, a freezing sixteen-hour bus ride to Santa Elena de Uairén (I took a warm jacket), a crossing into Brazil, another long bus ride to Manaus, then a four-day boat journey to the mouth of the Amazon.

The most interesting part of a trip down the Amazon is crossing the delta. This time I was going to Belém, a big city on the delta's southern edge. It's the main transportation link between the world of river travel that is the Amazon Basin and the network of highways and railroads in the rest of Brazil. This is where you finally see and feel the incredible scale and might of the river. Some channels in the delta are twenty miles wide, and islands can be the size of a small US state. Twice a day, tidal waves called *pororocas*, up to ten feet high, roll from the ocean for up to two hundred miles upstream. But the boat also passes through very narrow channels, with walls of dense rain forest on both sides. For the *caboclos*, people mostly of mixed Indian-Portuguese origin living along the river, these boats are the only connection to the outside world. So every time there's a wooden cabin on a riverbank, you see little half-naked kids jump into their tiny canoes and feverishly paddle to intercept the boat, hoping to exchange wood carvings and baskets of fruit for steel machetes and factory-made clothing.

Coastal Brazil south of the Amazon has an excellent highway system, so in just a few more days I was in Rio de Janeiro. I made a little detour to spend a night in Caraça, an old monastery hidden in a forested mountain valley in Minas Gerais state. Every night, local monks put out some minced meat, and maned wolves come out of the forest to eat it. These strange creatures look like giant red foxes with cheetah legs and guilty smiles on their faces.

In Rio I waited for Paolo, an old friend of mine. We had first met while hitchhiking in Patagonia ten years earlier. Paolo has all the skills and talents needed to be a good pirate, but he has a very boring job as a factory manager for a soft drinks company in São Paulo, so sometimes I succeed in talking him into one crazy adventure or another. Our previous raid was *para atraer a las chicas* (I'd rather not translate) in Uruguay. This time he took a month-long vacation to help me study caimans in the Pantanal.

He picked me up, and we drove inland, leaving the lush Atlantic rain forests and crossing the endless expanse of the Cerrado (savanna woodland), a country of small trees, tall grass, and huge red termite mounds. Driving in central Brazil is a lot like traveling around the US: good roads, well-developed infrastructure, lots of open space. But, unlike in the US, where you can never drive the way you like, in Brazil you enjoy the splendid feeling of absolute freedom. There are speed limits but no enforcement, and you don't have to stop at red lights unless there's considerable traffic.

After a few days we got to the Pantanal. Only two roads penetrate it. We followed the northern road, called Transpantaneira, until it ended at a river crossing in the tiny village of Porto Jofre. Beyond the river the roads were so bad that Paolo didn't want to use his truck on them. So we decided to rent horses. A full day of asking around and talking with ranch owners led us to a small cattle ranch downstream from the crossing.

The owner was a man in his sixties named Fernando. Like just about everybody in the Pantanal, he had an impressive mustache and an interesting biography. Many years earlier, he illegally emigrated to the United States, got a good job, married, and had two children. But his wife kept nagging him about not making enough money, so he decided to return to Brazil to start his own business. Within a week of coming back, he was tricked into investing all his money into some fake development scheme and found that he couldn't enter the US again for seven years (a routine punishment dispensed by US authorities to illegal immigrants). By the time his exile was over, his wife had obtained a divorce and married again. His children were grown up. And he found it difficult to adjust to city life again after seven years of working as a ranch hand. So he returned to Brazil for a second time and bought a small ranch, just a thousand acres. Land is cheap in the Pantanal because it's flooded for a few months each year, and only rare upland areas are available for grazing at that time. Bad grazing was what saved the wetland from development, making it the best place in South America to see wildlife.

We rented two horses from Fernando, set up camp on a wooden platform built in a tree by deer hunters, and started looking for good lakes to use in our study. Fernando told us which lakes were likely to break into small ponds as the drought continued. The rains hadn't arrived yet, so the land was very dry, and riding around from dawn till dusk was easy. We didn't even need a tent, as there were no mosquitoes at night—for mysterious reasons they preferred to fly in the heat of the day.

Unlike in Africa, seeing large animals in South America usually requires a lot of effort and luck. The open plains of the Pantanal are an exception. Every day we saw plenty of deer, capybaras, peccaries, and crab-eating foxes. Flocks of hyacinth macaws, the world's largest parakeets, gathered every evening at a night roost in the courtyard of Fernando's mansion. The horses were often startled by the smell of a jaguar or by a giant anaconda crossing our path.

Being surrounded by so much wildlife was a lot of fun. For example, we discovered that if you found a bunch of capuchin monkeys feeding in a cashew tree, stood underneath, and looked them straight in the eyes, they'd start throwing fruit at you—a nice way to obtain delicious fresh cashews.

One night we found a forest rat that had fallen into a dry well. We didn't have a rope, so we broke off a dead tree by ramming it with our truck, lowered it into the well, and picked up the rat as it clung to one of the roots. When the rat recovered a bit from stress and near-starvation, it became very beautiful, with golden-orange fur and glossy black porcupine-like quills. It was completely tame and lived with us for a few days.

Yacare caimans were everywhere. Lakes and even small puddles were chock-full of them. We chose six larger lakes, very shallow, with easily accessible shores, and began our work as caiman portrait artists.

I had a heavy stack of preprinted identification charts. Each chart had outlines of a typical yacare head and a caiman body as seen from both sides. Every time we saw a caiman "sing" with infrasound (which meant he was a male), we sketched his spots, scars, and other distinctive features on a chart and gave him a number. We also tried to get a few photos of him. The only catch was that we needed to see both sides of his head, since the markings on the left and right sides didn't match. So we had to run around part of the lake, or wade in, or launch my kayak to paddle around the animal. The caimans didn't pay much attention to our presence. They started roaring and headslapping at sunrise, then had a long siesta and did more "singing" in the evening. Their roars sounded more like barks. In fact, there is a small deer in Asia called a muntjac whose alarm calls sound exactly like yacare roars.

In just a week we portrayed eighty-eight adult male caimans and recorded at least three "songs" from almost all of them. Then we took a week-long break to drive around Mato Grosso Plateau and returned to the Pantanal just as the first rains started falling on hot afternoons.

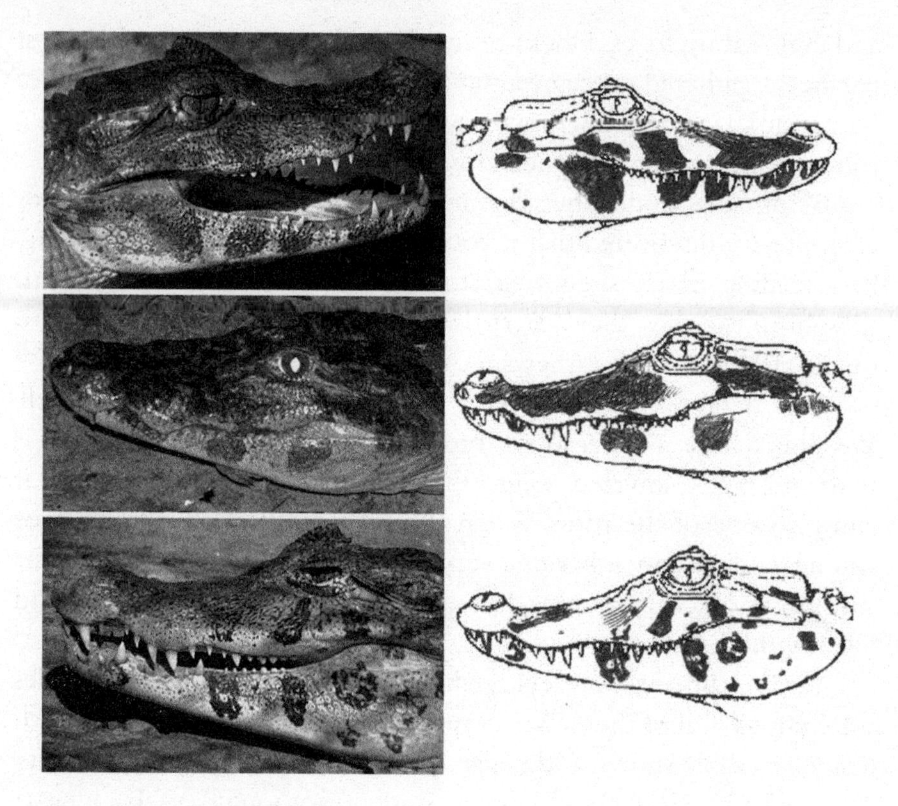

Each shower lasted for less than an hour. The country was rapidly changing from red and yellow to green, but the lakes were still shrinking. Four out of six of "our" lakes did exactly what we hoped for and broke into groups of small ponds. One lake didn't break into small enough fragments, and one was so shallow that it turned into a muddy puddle where caimans were stacked three layers deep, giant Amazonian leeches hanging from their bellies.

The air was getting hotter and more humid every day, but shrinking lakes were now surrounded by nice beaches, and we could swim any time we wanted. Some ponds were so full of fish that we had to push large catfish away to enter the water. Immense flocks of wading birds arrived at the Pantanal to feed on fish soup as the ponds kept drying out.

We managed to find most of our caimans and observed them for another week. Eventually, we got enough data (three "songs" recorded before the week-long gap, and three after it) for forty-four animals—a good sample size.

One day I rode far west, to the border of the national park. Only a small part of the Pantanal is officially protected in that park, but most Pantanal ranchers are extremely conscious about protecting the wildlife on their land. Recently, this grassroots conservation has begun to pay off for people living close to roads, as one ranch after another gets converted into a tourist lodge. But the more remote parts, like the area we were working in, still haven't seen a single tourist.

My mare lost a horseshoe, so I walked her to a nearby ranch looking for a smithy. The owner helped me "change tires," as he put it, and told me his life story as we were sipping hot maté (a drink made of a species of holly) on his porch. Of all the strange life stories I've heard there, this was the strangest.

Carl was born in Australia in the 1920s. His father was an Aborigine, his mother French. When World War II began, he was sent to North Africa as a military interpreter. He got captured by Rommel's troops and sent to Germany—one of very few people of Aborigine origin to become a prisoner of war. Since the Nazis considered him to be on the very bottom of their racial scale, they didn't bother sending him to a prison camp and instead gave him to a landowner as a slave worker. The landowner was a high-ranking SS officer, but that didn't prevent his daughter and Carl from falling madly in love. The girl apparently had total control over her father, for when he found out, he didn't send Carl to a death camp but allowed them to marry in secret. In 1945 the family escaped from Germany and made it to Paraguay. After his father-in-law, a wanted war criminal, died in the 1970s, Carl moved to Brazil with his wife, children, and grandchildren, went to Australia for a few months, was given a military pension, and bought the ranch. He was eighty-seven when I met him, his wife was long dead, and all his countless progeny lived in Australia. He didn't seem lonely and would ride around his property every day, supervising the farm crew.

It was time for me to go to Bolivia and look for my second study site. On the night before our departure it rained for six hours, and we spent two days getting the truck back to the Transpantaneira across twenty miles of mud. Paolo had planned to drive me to the border crossing. We stopped in the last big town, fifty miles from the border, because I needed to get an exit stamp in my passport and the police station was closed until the next morning. After we checked into a hotel, Paolo went to a local dancing club to pick up a girl. The girl he picked up had a boyfriend, so he barely escaped back to the hotel. At four in the morning we were awakened by a rock flying into our open window: the boyfriend and his numerous friends had gathered outside and wanted to talk to us. We had to jump into Paolo's truck and leave through the back gate. We drove to the border as fast as we could—and then found out that the city we'd just left was the only place where I could get that stamp.

Exit stamps are a big deal in South America. It used to be that anyone trying to enter a country without an exit stamp from the previous one was automatically considered a spy and thrown in jail. Recently, the continent has become a bit less paranoid, but it's still officially impossible to obtain an entry stamp without an exit one. And finding a place to get those stamps can be very difficult. In some cases it's easier to just cross all borders illegally, then go to your consulate in the last country on your itinerary and claim that your passport was stolen.

My Bolivian visa was expiring the next day, and Paolo had to get back to work soon after. I couldn't return to the city, so I decided to cross into Bolivia anyway. If there was trouble, I could simply bribe the official to get that entry stamp. Shouldn't be a problem, I thought.

14

Help from the Sky

Caiman yacare

Love always comes to an end, but so does life.
—Pablo Neruda

THE IMMIGRATION OFFICER on the Bolivian side of the border was sitting in a tiny room hidden in a narrow side street. He was also the town's only lawyer, real estate agent, wedding photographer, and notary public. When he noticed that I didn't have a Brazilian exit stamp, he flatly refused to issue an entry one. I explained that I couldn't go back to Brazil because my visa had expired and my Bolivian visa was about to, but he was adamant.

I needed that entry stamp. There were strikes and riots all over eastern Bolivia, so I could expect to run into checkpoints on every road. I put $40 in my passport and gave it to him again.

What happened next was like seeing a caiman fly. The officer looked inside the passport, sighed solemnly, stamped it, and gave it back to me *with the money*. Nothing I had ever seen or heard in that part of the world could prepare me for such a supernatural event. There had been rumors on travel forums about Chilean police not asking for bribes anymore, but nobody took them seriously; besides, Bolivia wasn't Chile. I stepped back into the street knowing with absolute certainty that I had just witnessed a miracle. What was going on? Was it the end of South America as I knew it? Was its entire way of life about to fall apart and disappear?

Deeply shaken, I got on a bus to the next town. Other passengers looked strange: blond, blue-eyed, and snub-nosed, dressed in long robes, men with beards, women with long braids. And they were speaking some really weird Russian. I introduced myself and soon learned that they were members of one of numerous Old Believer sects. Their ancestors left Russia in the late nineteenth century and established rural settlements in Brazil, Bolivia, and Alaska. Talking with them was fascinating: their Russian was frozen in time, having been transmitted orally for five generations. Their kids learned to read and write in Portuguese or Spanish in schools, but none of them had ever seen Cyrillic script. My own Russian suddenly felt full of Western words "imported" during the last hundred years. I tried to be very careful and not use them, especially when talking to children, because I didn't want to infect them with those new words and destroy the living fossil that was their beautiful Russian. You don't hear such rich language in Russia anymore. There are a few remote corners of the country where people still speak local dialects, but their speech is increasingly spoiled by the TV, where an ugly official version of Moscow dialect is used.

I had to wait for three days in the small town of San Ignacio de Velasco for the next bus. The most unforgettable experience there was the Mass, conducted on Saturday evening in a beautiful seventeenth-century Jesuit church. The electric lights around the building attracted a huge swarm of *Lethocerus gigas,* giant water bugs. They are the largest flying insects in the Americas, up to six inches long, with an excruciatingly painful sting that liquefies muscle tissue and sometimes leaves gaping never-healing ulcers. The bugs were buzzing around the church, crawling in layers on its walls, and bumping into people entering or leaving. The whole scene looked like a hellish upgrade of the Plague of Locusts.

The bus ride to Noel Kempff Mercado National Park took about thirty hours. By now it was raining daily, so the road was in sad condition. The driver swore not to do it again until the dry season, and stopped twenty miles short of the park. I had to haul my backpack, heavy with the inflatable kayak, paddles, and the pile of caiman identification charts, through sticky mud sometimes reaching up to my knees for a whole day to get there.

Since the idea to conduct this study of yacare caimans had come to me just two months earlier, I didn't have time to obtain a research permit for the park. So I visited it for a couple days as a tourist and then found a nice place in a remote marshland outside its southern limit, near the Brazilian border.

This park is a beautiful place, with waterfalls rushing down rocky escarpments, rain forests blanketing river valleys, and tallgrass savanna covering the uplands. Thanks to such diversity, the park and uninhabited borderlands south of it are teeming with wildlife.

The most common wildlife species are mosquitoes, chiggers, and sweat bees. Sweat bees don't sting, but they can be annoying when thousands of them cover you from head to toe on a hot day, especially if you are ticklish. And you have to be careful shaking them off, because once in a while there's a larger bee or a wasp mixed with them. As for chiggers, they become a problem only after you get back to civilization, by which time you have developed a habit of constantly scratching their bites (they can itch severely for up to a month), and these bites are usually in places you aren't supposed to scratch in public.

Eastern Bolivia is said to be the only place in the world where all five species of caimans can be found. But, except for two small black caimans inside the park, all I saw were yacares. Lots of them. Even working alone and having no horse, I catalogued and observed over eighty adult males in eight small lakes in just four days. It was raining most of the time, but there were still a few rainless hours every morning, and the caimans used them to "sing" every few minutes.

Other animals were also busy. Little, delicate pampas deer walked around with cat-sized fawns, white-lipped peccaries with striped piglets, and rheas (small South American ostriches) with similarly striped chicks. Opossums and giant anteaters were carrying cubs on their backs.

Then the big rains came. In just three days the entire valley was flooded. I didn't have a tent or a machete to build a shelter, so I slept first under my kayak, then in it, covering myself with banana leaves. I managed to keep my papers and camera dry, but everything else was soaked. Fortunately I didn't have to cook because there was no food left.

When the rain finally stopped, there was no dry land anywhere in sight. My caimans were still "singing," but finding them was extremely difficult. Everybody was on the move. Snakes, armadillos, and even sloths swam to high ground. Trees with water-dispersed seeds started fruiting, and immense flocks of macaws (there are eight species of them in Noel Kempff Mercado) descended on riverside forests to feed.

After three days of constant paddling up and down the floodplain in search of caimans, I gave up. I had enough data gathered before and after the flooding on twenty-six caimans, and that was more or less enough. I returned to the park office, and learned that the road out had been completely washed away, all bus drivers were on strike together with the rest of the province, and the landing strip was about to be flooded. It looked like I was going to be stuck there for a few weeks, if not months.

At least the telephone was still working. The only person I could ask for help was Jesus, the guy I met during my previous trip to Bolivia who owned a small plane. To my surprise, he remembered me. I asked if his plane was still operational. Jesus laughed and said he now owned three planes. He promised to fly me out in a few days for the cost of a bus ticket.

Then I called Carmen, and told her that I could be in her town in about a week. We hadn't seen each other for more than a year, but she didn't sound as happy as I hoped she would. She said I should call her when I got there, so I decided not to worry until I saw her.

Walking to my sleeping place in an old shed outside the village, I noticed an abandoned well and decided to check it for animals that could have fallen in. The well was filled with water almost to the rim, and inside was a small broad-snouted caiman, the first one I ever saw in the wild. It was two feet long and looked very similar to a yacare, but with jaws better adapted to cracking snail and turtle shells than for fishing. The caiman had probably spent a long time in the well and was so emaciated that it almost didn't move when I got it out. I named it Twiggy, put the poor creature in an empty gasoline tank, and fed it fish for three days. By the time I released it, it looked much better and was very agile. It even managed to bite my hand, but didn't leave a scratch. A year earlier a caiman less than half its size sliced my fingers like a food processor. Was Twiggy gentle on purpose? I'll never know.

While waiting for Jesus to save me, I tabulated my yacare data. The results were clear and straightforward. There was no change in the proportions of roars and headslaps in caiman "songs" before and after the changes in the size of their lakes.

I wasn't very happy about it, because it was a negative result, though it was going to make a nice chapter for my thesis anyway. But to get the thesis published, I'd have to turn each chapter into a separate paper, and submit them to journals one by one. A study with a negative result would be very difficult to publish by itself. I wasn't sure these data were worth three months of working in the field and crisscrossing the continent.

I didn't quite know what to make of this finding. Other animals have been shown to be able to change their signals in response to habitat changes. Even male jumping spiders can choose between two ways of impressing females. If they are on a solid substrate, they tap a seductive rhythm with their front legs. But if they are on a porous surface where vibrations don't spread well, they "dance" instead. How come the caimans couldn't adjust their "songs" to their habitat? Maybe my theory was wrong, and the differences between alligator populations that I had previously found had some other explanation? Or maybe the caimans didn't change their behavior immediately, but only after some time? I had to wait until the next alligator mating season to look for answers.

When Jesus landed, the airfield was so wet that I doubted we'd be able to take off. But he didn't seem worried. He said he was returning from a trip to Brazil. I didn't ask any questions.

The strike was still on, and all roads leading from the lowland part of the country up into the Andes were blocked by coca farmers protesting against trade restrictions on their produce. I asked Jesus if he could fly me straight to the Altiplano, the high plateau where most Bolivian cities are located. He didn't mind, and even agreed to drop me off in Carmen's town. We took on board another passenger, a stranded Israeli botanist.

Flying through the deep mountain valleys in heavy fog was a bit scary, but Jesus didn't care. As soon as we got out of the clouds and saw the bleak yellow hills of the plateau, he offered to teach me flying loops. It was so much fun! He also let me do the landing, my first one at high altitude. Only then did we remember about the Israeli in the backseat, and looked back expecting to see him half-dead with motion sickness. But he was peacefully sleeping. When we woke him up, he told us that he was a former paratrooper.

Before leaving, Jesus gave me a phone number of his friend in Colombia. "Carlos flies to Brazil every week or two," he said. "He might be able to give you a ride sometimes." By now I was sure I shouldn't ask any questions at all.

I called Carmen, and soon she showed up, beautiful as ever. She looked surprisingly adult, serious, even pious. The first thing she told me was, "I'm getting married next month."

I could only congratulate her. Her fiancé was a medical student, which sounded good enough. She brought me an alpaca sweater. I really appreciated it, for there was no place for warm clothes in my backpack, and nights on the Altiplano are usually freezing. She walked with me to a hotel and eventually agreed to stay for the night, but it wasn't a merry date. We both knew it was our last one.

The next day I got on a bus to Chile. It arrived at a border crossing on a frigid mountain pass at midnight, and we had to wait for the checkpoints to open in the morning. While we were waiting, one of the passengers, a lady from the lowlands, got so stressed by altitude sickness that she gave birth to a very small boy. For some reason this seems to happen almost automatically on trains, boats, and buses when I travel. I can now consider myself a reasonably experienced fake midwife. The important things to remember are to look like you know what you are doing, bark commands at everybody else, put medical jargon into anything you say, and let the process of childbirth take care of itself. This time I was lucky to have warm, albeit slightly rusty, water from the radiator, scissors, and some thread for tying off the cord. But I'm sure it would have ended equally well without my assistance.

As dawn broke, I saw a strange scene. Long queues of buses, cars, and trucks were waiting for the border checkpoints to open. On both sides of the road, just a hundred yards away, thousands of people were simply walking from one country to the other, carrying suitcases, backpacks, and burlap bags. These were happy locals who didn't have to worry about entry and exit stamps.

I didn't have that luxury because I had to cross four more borders to get back to Venezuela. So I waited patiently until eight o'clock, got the coveted stamp, and then had to wait for another hour while the happy parents tried to explain to Chilean border guards why their newborn child didn't have any documents.

Finally the bus descended from the Andes to the coastal highway. Just forty-eight hours after leaving the rain forest, I found myself in the hot, lifeless Atacama Desert, the driest place on Earth. I had to follow the coast and the Andes all the way to the north. The mating season of American and Orinoco crocodiles in Colombia and Venezuela was supposed to start in about two weeks.

15
Logging Observations
Crocodylus acutus

Keep a diary and someday it'll keep you.
—Mae West

CHILE IS THE MOST SCENIC COUNTRY IN SOUTH AMERICA, and one of the least visited. I could spend only a couple days there, so I rented a car in Arica and made a quick side trip to Lauca National Park, one of the best places to see Andean wildlife. The road from the park back to the coast is an almost unbroken descent from fourteen thousand feet to sea level and popular with mountain bikers. Just as I left the cold highlands and was crossing the groves of giant cacti at the upper edge of lifeless desert, I saw a bunch of bikers on the shoulder, trying to flag down my car.

I stopped. The bikers were from a tour group. The minivan that was supposed to follow them had disappeared, and they were too tired to pedal the last twenty miles to town. I managed to squeeze three girls and their bikes into the car and drove them to their tour agency. A few hours later I met the girls again on a bus to the Peruvian border. They were Brazilian journalists on their way to some conference in Ecuador. When we got to Arequipa, a large town full of beautiful colonial architecture, I managed to talk one girl, a very lively, bright-eyed brunette named Joanna, first into staying there with me for a day, and then into continuing our journey north together.

She probably didn't realize what she was signing up for. I had already been to all the famous tourist destinations of Peru, so this time I was planning to reach some places way off the tourist circuit. It was going to be a long and tiring journey, especially for Joanna, who was very girly, took her journalist status a bit too seriously, and attempted the impossible task of looking sexy and businesslike at the same time while backpacking.

Fortunately, the country's roads and bus system had improved a lot since my first visit there in 1995. Peru itself has changed almost beyond recognition. In the 1990s it was desperately poor, undeveloped, and dangerous, with guerilla armies lurking in the mountains and corrupt police who were almost worse than bandits. Now Peru looked well organized and relatively prosperous, and its people were well dressed, well fed, and generally happy. But crossing countless mountain ranges was still tiresome, especially since we both had long legs and couldn't fit comfortably into small rural buses. Climbing to a high pass would sometimes take our bus half a day. After five snow-covered passes, Joanna still looked very sexy, but nobody would call the poor girl businesslike anymore.

We got to Cajamarca, a city somewhat similar to the ancient Inca capital of Cuzco, but with one Inca-era building completely intact, and no tourists. By that time my friendship with Joanna was rapidly approaching a more interesting stage. We spent a long afternoon walking around the old city holding hands, kissing in dark archways, and making inappropriate jokes in museums, so I was looking forward to our return to the hotel. Suddenly, Joanna said, "Do you mind if I check my email?" We stopped in an Internet café, and she spent forty minutes typing something with impressive speed.

It didn't bother me as long as the rest of the evening took its course, but that wasn't an isolated event. It developed into a bizarre pattern. Every time we were about to go to bed, or were out of our hotel room in the morning, she'd get online and spend a long time typing. She refused to tell me what she was typing, but once I looked over her shoulder and saw my name in almost every line. At that point she had to tell. She confessed that she was writing a column for a women's journal back in Brazil. Coming up daily with so much text wasn't easy, so she was describing our relationship to her readers in very minute detail. I didn't mind (it's not every day that you can read an honest description of yourself as a lover), but why mention my full name?

"Come on," she said, "consider it free advertising!" I tried to ignore this, but couldn't, because in our most intimate moments I'd remember the relevant descriptions from her column and start laughing. I decided to get back at her by describing her in detail in my LiveJournal blog. Of course, I didn't use her real name. No free advertising from me!

LiveJournal is where much of Russian-language intellectual life flourishes. For millions of expats it's a convenient way to communicate with their friends around the globe. For those still in Russia, blogging is the easiest way to express themselves in a country where the government censors all other media and the book market is in a pathetic state due to the traditional rejection of the concept of copyright. The ruling junta understands the importance of this new media and hires hundreds of people to find blog posts that are critical of the government, or pro-Western, or simply too intelligent, and flood them with pro-government comments. The scum recruited for that job are poorly educated and often barely literate. People usually ignore these interruptions, but sometimes the war of ideas gets pretty intense.

The Russian part of LiveJournal has tens of millions of subscribers, so the American creators of this blogging service have sold it to a Russian company. Since I was writing erotic pieces about Joanna, I decided to use the growing popularity of my blog for my crocodile research. I knew that in a few months I was to start a difficult, labor-intensive, and expensive study of Nile crocodiles. There was no way I could do it alone. I needed volunteers, preferably people capable of sharing the expenses. So I posted an ad in my blog. I don't remember the exact wording, but it was an honest ad, reminiscent of the famous 1914 job announcement for Ernest Shackleton's South Pole expedition:

> "MEN WANTED for hazardous journey, small wages, bitter cold, long months of complete darkness, constant danger, safe return doubtful, honor and recognition in case of success."

Of course, I couldn't promise any honor or recognition.

Eventually we got to Ecuador. I spent my last couple of days with Joanna in Quito, and continued to Colombia. I wondered what she was writing about after I had left, but her journal wasn't available online.

Of all South American countries I visited that year, Colombia had changed the most—even more than Peru. In 1995 it was in the middle of a civil war. Travelers were routinely mugged, kidnapped, or killed. Nothing worked. Now it was much better in most places, although some towns were still very poor, with dirt streets and a largely unemployed population. The civil war was rapidly coming to an end, and the country was relatively safe. Only once did I almost get robbed. It was in Rio Ñambi Nature Reserve, known as the wettest place in the Americas—it has rained there every day since the first records by the conquistadores. I left my backpack under a tree while walking around, and when I came back, a half-grown spectacled bear was chewing on it. I didn't want my kayak to get torn, so I tried to chase the bear away—and it tried to take the backpack with it as it ran away but got stuck in the dense undergrowth, dropped the backpack, and climbed a tree. It was a great photo op.

When I was within a couple days from Bogotá, I called Carlos, the private pilot that Jesus had recommended. Carlos was much more reserved than Jesus and didn't seem too enthusiastic about taking a stranger on board, but eventually he agreed to drop me off in Leticia on his way to Brazil, and then pick me up on his way back. His habit of snorting cocaine while piloting the plane over the eastern Andes was a bit unnerving, but he refused to let me take over.

Colombia owns the northwestern corner of the Amazon Basin. Much of that corner was still under the control of FARC guerillas, but the area around Leticia, the country's only port on the great river, was said to be safe. Carlos landed on a private airstrip twenty miles from the town, so I didn't have to find a way to get out of Leticia and into the forest: I just paddled my kayak for a few miles upstream and built a shelter with bamboo stems and banana leaves. Then I made myself a bow and a few arrows with palm frond tips for arrowheads. This was all I needed to get small fish from forest ponds and streams.

One large catfish I shot had little white wormlike things stuck in its gills. I looked closely and was excited to see that these were the candiru fishes, the most feared inhabitants of the Amazon.

One of the first things you hear almost anywhere in tropical South America is a colorful description of the candiru's terrifying habits. The fish is said to be attracted to the smell of urine. If you pee in a river, or sometimes even wade into the water, a candiru will squeeze itself into your urethra, often after swimming up the stream of urine, and anchor itself inside by spreading the sharp spines in its tiny fins. If you don't die from shock, you'll need a major surgery to have the parasite removed. Some local people are so afraid of the candiru that they never get close to any body of water at all.

This is all pure fiction. The candiru is one of many small species of catfish that lives under the gills of larger fish. It's not attracted to urine (there have been plenty of experiments), and cannot swim up a stream of urine (that's physically impossible). There has never been a documented case of a candiru being stuck in someone's urethra. One such claim from 1997 was a fake, although there are still lots of gory photos of the supposed surgery on the Internet. Over the last two hundred years there have been a handful of cases when candirus entered human vaginas, but the fish didn't do any damage and were easily removed by hand.

I spent two days in the forest looking for caimans, but they were scarce and extremely shy. It was obvious that they were still hunted a lot, despite being the less-valued spectacled caimans. I didn't get any interesting observations. Soon Carlos arrived from Brazil and flew me back to Bogotá. I ran straight to the bus terminal and managed to escape the city before the pre-Christmas rush, when you can easily get stuck for days in mile-long ticket queues.

Soon I was on the Caribbean coast, and headed straight for Salamanca National Park, a large area of mangroves near the city of Barranquilla. The park has one of the few remaining populations of the American crocodile in South America. I already had a lot of data on this species from Florida and the Dominican Republic, but I wanted to know if there were any geographical differences.

Five days later I knew the answer. Quiet, cough-like roars of Colombian crocodiles sounded exactly like those I'd already heard, and, just like in Florida, there were approximately seven headslaps for every roar. In the Dominican Republic I recorded only headslaps, but maybe I didn't wait long enough.

My theory predicted that the American crocodile, a species of mangrove lagoons and river estuaries, should use a lot more headslaps than roars. I was happy with the new data, and decided that I deserved a two-day vacation. I had an old dream that I could now fulfill.

16
Sadness and Hope
Crocodylus intermedius

The World is a book, and those who do not travel read only one page.
—St. Augustine

MY FIRST CHILDHOOD DREAM WAS VERY PRECISE and well formulated. I wanted to go to South America as a traveling naturalist. The dream mostly came from reading the wonderful accounts by Darwin, Arkady Fiedler, and Durrell (not much else was available), but also from realizing that no part of the planet was more different from my hometown. Moscow in the so-called "stagnation period," just before perestroika, was a boring place.

I soon learned that talking about my dream at school wasn't a good idea. The first time I told someone about it I got referred to the district psychiatrist. The second time I almost got my mother in trouble: discussing foreign travel, especially to any "America," wasn't a welcome subject in the Soviet Union. The lucky few who got a chance to see another country had to stay in a group and be accompanied by a KGB overseer at every step. South America seemed as distant as the moons of Jupiter.

By the time I was out of school, I realized that I'd have to leave the country illegally, and even did a practice run one summer, crossing the border from Tajikistan (still a part of the empire) to Afghanistan and then to Pakistan. I was planning to do it again, this time one-way, but in 1991 the regime fell. I made some money writing nature guides and working in Europe and the Middle East, and within four years was able to go to South America for six months.

I flew to Nicaragua, spent two weeks in Costa Rica, failed to get a visa to Panama, and took a flight to Caracas. The plane was supposed to land in Barranquilla first, so I was glued to the window, waiting for the Colombian coast to appear. And appear it did. Below was a walled city straight out of a fairy tale. It looked a bit like the castle at Disneyland but much more beautiful, laid out in the shape of a flower with bastions as petals, with cathedrals and tiled roofs squeezed inside. It was Cartagena de Indias, one of the oldest European settlements on the continent. It disappeared from view within a few minutes, but I promised myself to see it up close someday.

Now, seventeen years later, I was within a short bus ride from there and decided to go to Cartagena for Christmas. Much of the city is modern, industrial, and rather unattractive, but once inside the walls you find yourself in one of the prettiest colonial towns in the New World. It has more flower boxes per window than any other place outside England. It has horse-drawn carriages for taxis. It even seems to have more pretty girls per capita than Chile, which I thought was impossible.

Most of those girls were parading the streets accompanied by their older relatives, but in the evening, when I walked into a small café, I saw a beautiful young lady sitting all alone at the corner table. She looked very slender and delicate, even fragile. I asked if I could join her, and she smiled. We talked for some time, then spent a few hours walking around and watching the streetlights being lit. Her name was Soledad; she was very warm and charming but seemed strangely sad. Later we found a restaurant, and as soon as the waiter brought us the menu, she said:

"Before you pay for my dinner, I have to tell you something about myself."

"Sure," I nodded, trying not to look worried. I have a rather colorful imagination, which immediately started painting pictures I'd rather not describe. The pause she made was long enough.

"I'm a prostitute," she finally said.

I don't know what my face looked like. She quickly added:

"It doesn't mean anything tonight. I like you and don't want to be alone. Please don't leave. You don't owe me anything."

I've never paid for sex in my life, but I used to know a few prostitutes. One of many jobs I had in my first two years in the States was pizza delivery. I lived in Berkeley at the time, and there were lots of student girls in town who were making their living that way. They were the best customers, because they'd usually give me a twenty from behind a door and never wait for the change. Sometimes they asked me to drive to a store to buy this or that, so eventually I became almost friends with a few of them. They were nice girls, but I always refused their offers of a pro bono.

This time it was different. The girl was so sincere and lonely. The town was such a magical place. I looked at her and suddenly understood what she was trying not to show. It was desperation. How could I leave?

There are no secrets left untold in the age of the Internet. A few days after I left Cartagena, she wrote me a long email. I will quote only one paragraph. "To love" and "to want" are the same word in Spanish:

"You must be thinking I'm crazy. Maybe I am. We have a belief here: if a prostitute doesn't have sex just for love at least once a year, she will forever lose the ability to love. I hadn't met anyone I liked this year, so when you showed up and I realized that I loved you I thought God had sent you."

As if this wasn't sad enough, I got a second email the same day. I was in Tayrona National Park, resting on a beach after climbing Sierra Nevada de Santa Marta, the world's second highest coastal mountain range. I was very tired and my clothes were dirty. It was New Year's Eve. Backpackers frolicking around kept making jokes about my huge backpack. They didn't know there was a kayak inside, and not much else.

The message was from the wife of Jesus, my Bolivian friend who saved me from being stuck for months in the flooded jungle. It was just one sentence, sent to all his email contacts. Jesus was dead: his plane had been shot down over the jungle by the Brazilian Air Force. They didn't bother to find the wreckage and recover his body.

That was just too much. I couldn't stand happy faces around me anymore. I needed to get back to work.

I hitchhiked east along the coast, crossing the driest part of Colombia. It's almost a desert, dotted with cacti and never-rotting piles of garbage. The most famous place there is the small, dirty town of Maicao near Riohacha, which has supposedly been described as "Macondo" in *One Hundred Years of Solitude*. From Maicao it's just a two-hour walk to the Venezuelan border.

Venezuela had been the richest country in South America during the first oil boom of the 1960s; now it was one of the poorest. Every time you got into a conversation, it was about Hugo Chavez. People driving private cars mostly hated him; those riding cheap buses mostly worshipped him. But there were many exceptions. I tried to stay open-minded, but after a few weeks I had to admit that the country wasn't doing well. There were thousands of oil rigs all over the northwest, all pumping, but the roads were in poor shape, city slums looked worse than in Rio, and many small businesses were closed.

I made a week-long stop at the coast to look at the easternmost South American population of the American crocodile in Morrocoy National Park. The park has nice mangroves, but it's mostly famous for huge flocks of red flamingos and scarlet ibises. The results of crocodile observations here were virtually the same as in Colombia and Florida: one roar per eight headslaps. This species seemed amazingly consistent in its "song" composition throughout its range.

Crossing Cordillera de Mérida, the Venezuelan part of the High Andes, I stopped for a day to rent a paraglider. Spiraling down through the clouds from a snow-clad cliff two thousand feet high, I almost collided with a particularly inquisitive condor. That was my other childhood dream: to soar in some Andean canyon side by side with condors.

Beyond the mountains were four hundred miles of flat savanna known as *llanos*, crossed by the Orinoco River. Just a century earlier the llanos were inhabited by thousands of Orinoco crocodiles. They were the largest crocodilians in the New World, often exceeding sixteen feet in length. In just twenty-five years, from 1945 to 1970, they were almost completely wiped out by hunters. Today, this species would probably be extinct if John Thorbjarnarson hadn't started a captive breeding program. He also worked hard to persuade local landowners to protect crocodiles on their ranches. Venezuelan conservationists joined the cause and eventually reintroduced crocodiles to many places in the llanos. A national park was established, although it never got effective protection. There's still some poaching going on, but there are hundreds of Orinoco crocodiles in Venezuela now and a few dozen in Colombia.

For me this species was of particular interest. Genetically it's so close to the American crocodile that some experts consider it a subspecies. It differs in being larger, with slightly stronger armor and a very narrow snout. Unlike the American crocodile, which mostly inhabits brackish water, it lives only in freshwater. It prefers rivers, though in the dry season (which is also the mating season) many crocodiles are found in oxbow lakes and even small ponds. So my theory predicted that despite being so closely related to American crocodiles, they should use more roars and fewer headslaps.

I started with a visit to a breeding center at a private ranch called Hato Masaguaral. There were two croc pairs there. Both females had already dug their nests (this is one of the few crocodile species that nest in holes) and were extremely protective of them. If I got too close to the fence, they'd hiss and roar and ram the chicken wire. The males would remain in the water nearby, splashing with their huge tails and headslapping. But the males were still "singing" at dawn. Unlike the soft "coughs" of American crocodiles, their roars could be heard from half a mile away. The sequence was usually the same: infrasound vibration, roar, headslap.

John told me that the best place to look for them in the wild was Rio Capanaparo, a tributary of Orinoco. It was crossed by two highways less than a hundred miles apart, so all I had to do was take a bus to the upper bridge, float down in my kayak looking for crocodiles, then flag down another bus at the second bridge.

The crocs weren't easy to find: decades of hunting had taught them to stay away from the main channel. Spotlighting didn't work well because there were hundreds of spectacled caimans everywhere. You can sometimes tell the eyes of a crocodile from the eyes of a caiman (caiman eyes are raised above the head, so the distance between the eye and its reflection is larger when the animal is floating in the water), but it's difficult and you can never be sure, so you still have to paddle closer to each pair of eyes to check them out.

Eventually, I found two huge, formidable-looking males in two oxbow lakes separated by a narrow isthmus. This was perfect: I could watch them both at the same time. I didn't want to stay there for too long. The mating season was almost over, and besides, I didn't bring any food, and the water was too murky for bow fishing. Both crocodiles spent most of the night slowly patrolling their lakes, but in late morning they'd hide from the heat in huge burrows they had dug in the steep banks.

Narrow jaws of some crocodilian species are usually considered a sign of feeding mostly on fish. Crocodiles typically catch fish by a sudden sidewise jerk of the head, combined with slapping the jaws. The narrower the jaws, the less the water resistance, so the move can be done faster. Broad jaws are supposed to be better for cracking turtle and snail shells. But the difference in diet is not that big: fish is the main prey for all known crocodiles, and given a chance even the most narrow-snouted species would catch anything that moves. The two males I was watching often tried to sneak up on egrets or capybaras, and one of them almost caught a deer that came to drink.

Neither of the two lakes had any caimans, while every other pond in the savanna was full of them. Was it because the crocodiles were better at fishing, or because they'd attack caimans? I soon got the answer. One night a five-foot-long caiman was careless enough to walk into one of the lakes, probably from some drying pool. It was visibly exhausted. As soon as it floated away from the shore, the resident crocodile attacked it, seizing it by the chest and shaking it violently. The caiman managed to wrestle itself free, and tried to run up the bank of the lake, but the crocodile caught it by the tail and dragged it back. They moved to the far side of the lake where I couldn't see them well in the moonlight, and went quiet after a few minutes of splashing. By dawn, all that was left of the caiman was its head. A few yellow-headed vultures came to feast on it, and one of them was promptly caught by the crocodile. I suspect that Orinoco crocodiles regularly hunt caimans, but it has never been studied.

Strangely, there are no crocodiles in the Amazon Basin. Their niche there is probably occupied by the black caiman, which, in turn, doesn't live in the Orinoco. How this separation works is unclear, because there's a channel connecting the two river systems (it changes the direction of its flow depending on the season).

I had two weeks left before my flight home. Spectacled caimans didn't "sing" at that time, and the only other crocodilians in Venezuela were the elusive dwarf caimans. I didn't expect them to be "singing" because I knew that in Guyana they do it in August. But there was nothing else to do, so I decided to look for them. I floated down to the second bridge and took a bus south, across the mighty Orinoco.

17
Natural Selection
Paleosuchus trigonatus

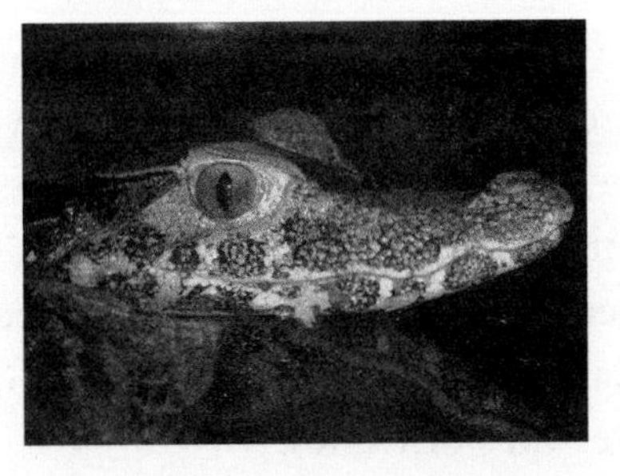

Travel only with thy equals or thy betters; if there are none, travel alone.
—The Dhammapada

SOUTHERN VENEZUELA IS UNLIKE ANY OTHER PART OF OUR PLANET. It's a rolling plain covered with shortgrass savanna in the east and dense rain forest in the west. Above the plain rise more than a hundred isolated mountains of red and black quartzite with vertical sides and usually flat tops. They are called *tepuis*, "houses of gods" in the local Pemon language. The tepuis are among the world's most ancient geological formations, the remnants of an elevated plateau that took two billion years to be mostly eroded away. Auyán Tepui is the most famous because the world's tallest waterfall, Salto Ángel, falls from its edge. The larger tepuis, such as Roraima, are islands of colder climates, with lots of endemic plants and animals on their summits. Some of the smaller ones have never been explored.

The southwestern part of this region is in the Rio Orinoco catchment. Thanks to a series of cataracts on the Orinoco just above the small frontier town of Puerto Ayacucho, its upper reaches are sparsely populated and relatively pristine. You can get by road to a small settlement above the cataracts, but from there boat traffic is scarce and roads absent.

Tourism in the area had mostly died out; all travel agencies in Puerto Ayacucho were closed. I waited for two days in a semi-abandoned tourist camp at the cataracts until there was a speedboat to yet another semi-abandoned camp, located on a small river near the four-thousand-foot Autana Tepui. Like many small tepuis, Autana resembles a colossal tree stump, but it's unique because a huge cave runs all the way through it, like a tunnel.

The river itself was too fast for caimans, but small streams running into it were just the right habitat. The ones flowing from the tepui side were clear; the ones emerging from the forest on the other side were blackwater streams, dark like very strong tea. Literature suggests that clearwater streams should be inhabited by Cuvier's dwarf caimans, and the blackwater ones by Schneider's dwarf caimans. But I found only the latter on both sides.

This species is slightly larger than Cuvier's, which I had observed in Guyana. It has the strongest armor of all crocodilians, a triangular head, and a very muscular tail—supposedly adaptations to life in fast streams. Recently it was discovered that these little-known caimans prefer to nest near termite mounds. Termites build complex ventilation systems, with warm air constantly rising from underground. The caimans apparently use that warm air as a heat source for the developing eggs. There's little sunlight in dense forests where they live.

The caimans I found had burrows under the overhanging banks of the streams. The best time to look for them was early in the morning, when they'd bask on snags for a few minutes before disappearing in the burrows, and at dusk, when they emerged on the surface. I tried to follow them as they went to hunt, but didn't see much. There wasn't enough moonlight in the forest and I couldn't use artificial light, except for weak red light from my small headlamp. White light made them dive immediately. Red light barely allowed me to see their eyes from about a hundred feet—the closest I could usually get to the little caimans without spooking them.

It's said that dwarf caimans often hunt on land and can walk for miles every night in search of food. But the ones I followed didn't do that. They swam upstream until they reached the uppermost stretches of their creeks. The water levels were low, so these creeks looked more like chains of small pools. This is where the caimans would hunt. On the few occasions when I was able to see what they were eating, it was small fish and once a dragonfly larva.

It was so nice to return to the camp every morning after a night spent crawling through vine tangles and spiderwebs along those streams. There was a gorgeous sandy beach, and I'd jump in the big river and soak there for most of the day, extracting thorns and letting little fish clean my skin. The water was so murky that only the top layer, a couple inches deep, was very warm, and the rest was wonderfully cool. Giant damselflies fluttered above the water. They were black with yellow wing tips and heads, so each looked like five small butterflies flying together. Once I came back from the forest earlier, before dawn, and saw four bright golden eyes reflecting my flashlight from underwater. They belonged to a pair of beautifully patterned stingrays dancing in the shallows.

The reason I came back so early that particular morning was that the caiman I had been following suddenly stopped hunting around midnight, returned to its burrow, floated just outside the entrance for two hours, then raised its head and tail and roared. An hour later it raised its head and tail again, but this time it made a headslap, and soon disappeared underground.

Unfortunately, I had to leave before I could get more data. There were no boats going downstream, so I had to paddle like crazy for a whole day to get back to the road. Then I hitchhiked to Puerto Ayacucho and took a very slow bus to Ciudad Bolívar, the largest city on the Orinoco.

Ciudad Bolívar has a colorful historic district and great views of the river. When I was there eleven years earlier, in 1995, a lot of people were suffering from river blindness, a disease transmitted by tiny gnats, but it has been eradicated since. I was in a hurry, so I took another bus, then yet another, and ended up in Tucupita, a city in the Orinoco Delta. All businesses were closed there. Even the gas station would open for only two hours a day. I don't know why nobody seemed to be working in Venezuela, but it reminded me of the 1980s in the USSR, when people would show up for work and do nothing all day because they didn't care anymore.

I managed to find a boat to go deeper into the delta and have another try at spectacled caimans, but by the time I got to a place that had them, the boatmen had already told me that the mating season was in March. I watched the caimans until the next day just to make sure, returned to the town, and took a bus to Caracas. Like the rest of the country, the entire city was plastered over with portraits of Chavez in the Nazi-Soviet style of official adoration, usually surrounded by workers and farmers, with factory smokestacks and cornfields in the background.

A short flight over the Caribbean, and I was finally home. I always used my blog to find someone to stay in my Miami apartment while I was away. It was just thirty minutes from the beach; I let people use it and my car for the amount of rent I was paying for the place. But this time the family who had spent their vacation in my apartment had left almost a month before my return. So there was a large anthill in the bathroom, with ant highways branching out to all rooms, and four tiers of cobwebs full of dead ants were hanging above the anthill. Window frames were covered with green algae, the fridge was moldy inside, the mailbox was full of paper gruel, also moldy. The fruit flies in the trash bin were extinct due to the lack of food, so all the geckos had moved out and cockroaches ruled the place. A wolf spider had settled under the bed, and a box of sugar left on the dinner table had turned into a puddle from excessive humidity. Outside, the windshield-washer pipe in my car was completely rotten, and a bunch of "magic" *Psilocybe* mushrooms were growing on the front lawn.

I love living in the tropics!

I have to admit that I spent most of the first few days reading LiveJournal blogs, posting photos, and exchanging comments. There was one girl whose Russian-language blog I particularly enjoyed reading. She had beautiful style and a great sense of humor. I wrote a couple appreciative comments, and soon we became online friends. She lived in Tennessee, just a day's drive away. We decided that I'd stop by during my next alligator trip. Her name was Nastia, short for Anastasiia.

But for now I had too much work to do. I had less than a month before going to Mexico, where Morelet's crocodiles were supposed to be mating. I had to plan the year-long African expedition and begin interviewing volunteers for it.

A lot of people had replied to my ads for an assistant, but most of them wrote things like "This is my life's dream, but I'm just too busy right now." I knew very well that they'd never go anywhere; they'd just keep postponing it until they got too old. One very nice British girl—a biologist!—almost signed up, but eventually decided to become a truck driver for an overland tours company instead. There were a few people living in Russia who wanted to go, but I couldn't take them because it would be close to impossible to get them South African visas. I tried to talk Nastia into joining, but she had already bought tickets to visit her family in Ukraine in June and had to be back at her university by mid-August.

So I was down to four candidates, and invited them all for interviews. One lady named Shirley lived in New York, so we talked on the phone. She sounded OK, and I decided to meet her in person later. I was planning to be in New York in a few months. One girl came to Miami for the interview—and within five minutes we both knew it wasn't going to work. She was way too normal for such an extreme adventure.

The remaining applicants were a young couple, Sarit and Alex. They were both born in the former Soviet Union and had come to the US as teenagers. Alex was a talented engineer; Sarit was trained as a painter. A few years earlier Alex successfully sold his start-up, and, being fed up with the standard middle-class way of life, he and Sarit went traveling with no intention of ever coming back to their boring jobs. I was lucky to catch them during their short visit to their families. They had just returned from a long trip around Central America in a "hippie bus."

They were staying with Alex's parents in Tamarac thirty miles north of Miami, so meeting them in person was easy. As soon as we met I knew they were perfect candidates, just the kind of people I needed: free from any obligations, experienced travelers, pathologically prone to risk-taking behavior, and capable of sharing the expenses. Alex was really good at fixing all kinds of things and generally very smart. Sarit was extremely social, so I knew she could do all the talking. You have to do a lot of talking to get anywhere in Africa. They both spoke Russian, so we could communicate without others understanding.

As nice as they seemed, I wasn't going to take them on a year-long trip to Africa without some personal compatibility testing. So we designed a simple plan for that spring and summer. They were about to leave for Honduras. I would meet them there for a few days after my research in Mexico and Guatemala. Then I'd come back to Florida and spend April and May studying alligators. My new friends would return from Honduras and I would then drive them to New York (the cheapest flights to Africa were from there). We'd pick up Shirley, travel together in Canada for two weeks, and return to New York. Sarit's mother lived there, so I could leave my stuff and car at her place. In late June I had to fly to Denver, meet with my mother, who would come from Russia, and drive her around the West for three weeks—it was something I had promised her earlier. As soon as I was back in New York, my volunteers and I could fly to Johannesburg, rent a small car, drive to Cape Town, lease a truck, and drive fifteen hundred miles to the Caprivi Strip, a narrow band of Namibian territory sandwiched between Angola, Botswana, and Zambia. After that it was going to get complicated.

18
Politics of Extinction
Crocodylus moreletii

Never ask God to give you anything;
ask Him to put you where things are.
—Mayan proverb

THERE ARE FOUR SPECIES OF CROCODILES IN THE NEW WORLD, and their origins are surprising. All four look quite different, and anybody could easily learn to tell them apart. But genetically they are very closely related and hybridize freely, producing fertile offspring. The fossil record and molecular studies suggest that at some time between three and seven million years ago, a single female crocodile swam from Africa to the West Indies, and her descendants began colonizing the Americas—until then the exclusive domain of the alligator family (alligators and caimans). The ocean at that time was almost as wide as today, so crossing it was an impressive feat, even for an animal that can fast for a year. Females can also store sperm for over two years, so laying fertilized eggs after the journey wasn't a problem.

Once in the New World, crocodiles split into four species: the relatively widespread, mostly coastal American crocodile and three freshwater species with more local distributions. Interestingly, the evolution of these three freshwater species went in somewhat opposite directions. Orinoco crocs became very large and narrow-snouted.

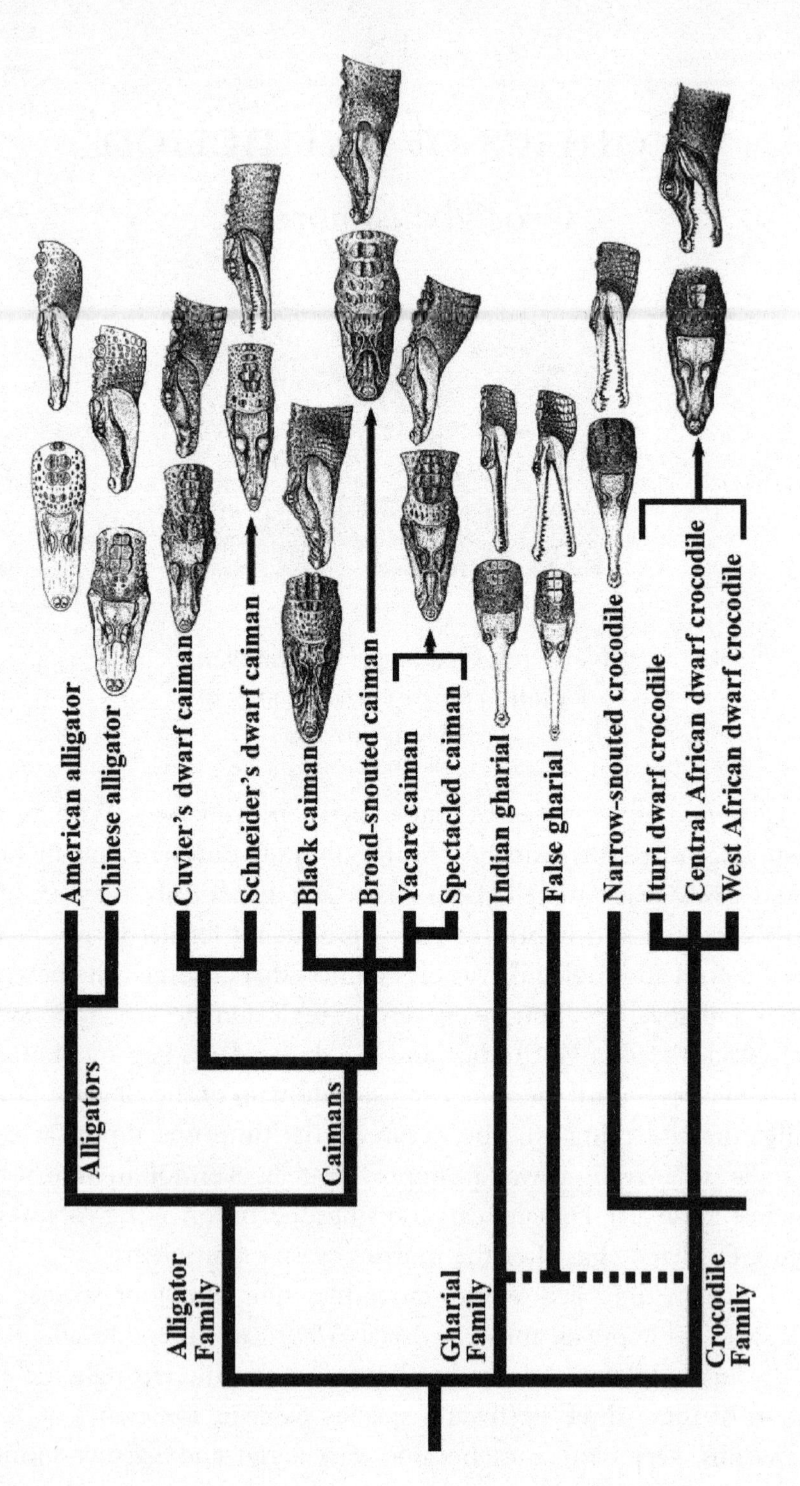
American alligator
Chinese alligator
Cuvier's dwarf caiman
Schneider's dwarf caiman
Black caiman
Broad-snouted caiman
Yacare caiman
Spectacled caiman
Indian gharial
False gharial
Narrow-snouted crocodile
Ituri dwarf crocodile
Central African dwarf crocodile
West African dwarf crocodile
Alligators
Caimans
Alligator Family
Gharial Family
Crocodile Family

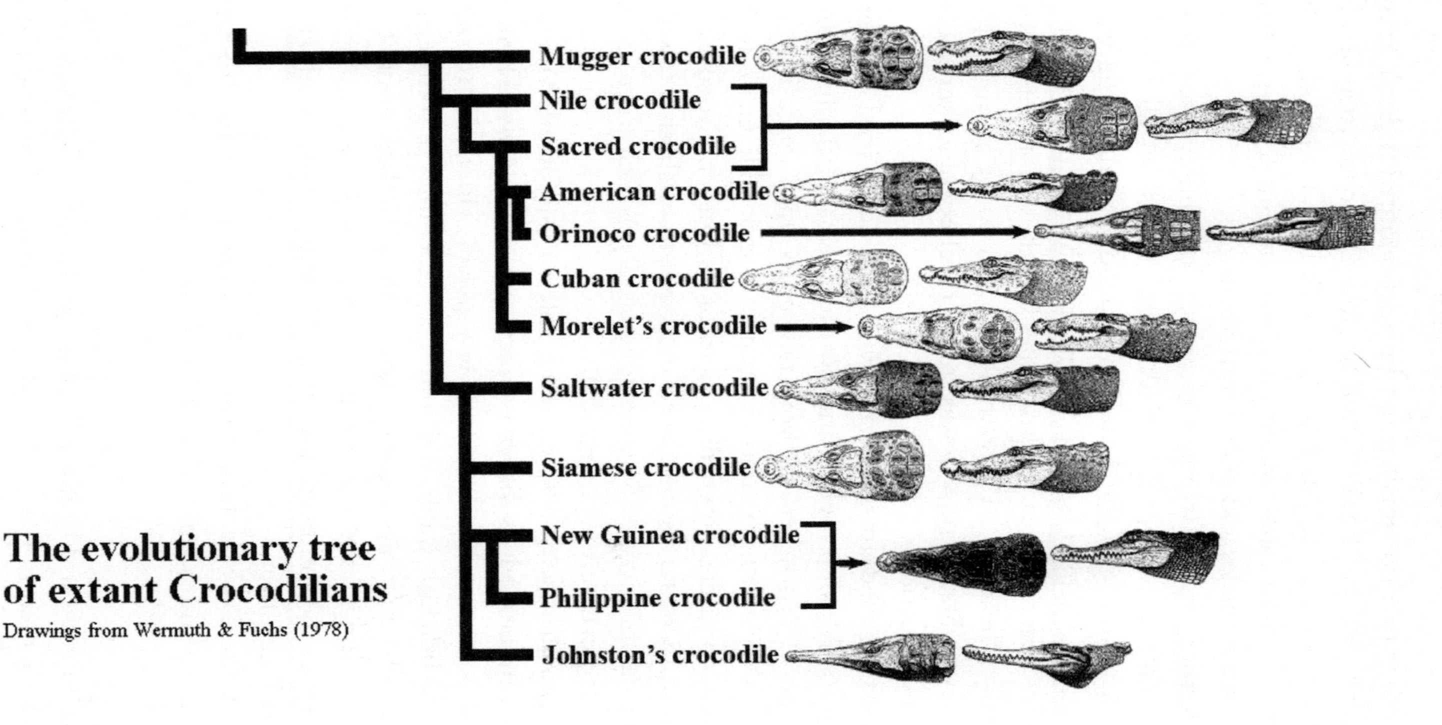

The evolutionary tree of extant Crocodilians

Drawings from Wermuth & Fuchs (1978)

The other two got smaller and broader-snouted. I don't know why this happened, but I guess it was because Orinoco crocodiles evolved in large rivers and oxbow lakes, very rich in fish, while the other two adapted to freshwater swamps, marshes, and small lakes, where they had to hunt turtles and crabs more often.

Despite the different appearance of these three freshwater species, my theory predicted that the "songs" of all three should differ from those of the American crocodile in the same way. Since their aquatic habitat is more fragmented than mangrove lagoons and estuaries where American crocodiles live, they should use more roars and fewer headslaps. I already knew that to be true for the Orinoco crocodile, so now I was going to check another species: the Morelet's crocodile.

This is a medium-sized croc with a Gallic bridged nose that lives in eastern Mexico, Belize, and northern Guatemala. Its skins are of good quality, so it became very rare in its small range in the 1970s. Now it's recovering, thanks to numerous crocodile farms in Mexico that have brought down the prices of its leather. Crocodiles escaping from those farms have even managed to establish a few colonies on the Pacific coast of Mexico, where only American crocodiles and spectacled caimans occur naturally.

It's unclear why, despite being able to interbreed, Morelet's and American crocodiles haven't mixed up completely. Their habitat separation is not absolute. In a few places Morelet's crocodiles are found in brackish water, and American crocodiles of southern Mexico often live in rivers and large lakes. I thought there could be differences in "songs," and was going to check this. I was now on my way to Chiapas, the southernmost state of Mexico, where both crocodiles were still common.

But there was a little private business that I had to take care of first. So instead of flying directly to Chiapas, I took a plane to Guatemala.

Before moving to Florida, I lived in the West for eight years—first in California, then in Colorado and New Mexico. The southern border was never far away, so I often traveled to Mexico. In 2003 I decided to get as far as the road goes, and drove from Boulder to Panama and back. At that time I worked every summer as a field technician with research teams studying the distribution of rodent-borne diseases, such as plague and hantavirus, so I was particularly interested in rodents. In northern Guatemala I decided to explore a remote mountain range called Sierra de los Cuchumatanes. It's a plateau up to 13,000 feet high, not unlike the High Andes of South America. The towns in its foothills are famous for strange pre-Colombian cults and carnivals, often involving bloody contests. But what interested me the most was its northeastern slope, where montane forests still survived.

In 1975, a new species of rodent—the Maya mouse—was described from the upper edge of these forests. It had dense black fur and a long nose, fed mostly on insects, and could survive only in mature forests with a thick layer of leaf litter on the ground. Such a thick layer forms only in cold places, where leaves don't decompose too rapidly, so the mouse lived in an unusually narrow elevation range, from 9,700 to 10,000 feet.

After looking for such forests for a few days I realized that they all had been logged. There was still plenty of forest downhill, but only pastures above 9,600 feet. Although the entire range was a nature reserve, there was no real protection, so logging was rampant. Finally, I found a tiny patch of forest growing on a steep slope, with one huge oak and a bunch of smaller oaks, pines, and rare Guatemalan firs. The grove was surrounded with fresh stumps on all sides. It was Friday evening; the loggers were clearly going to finish their job next Monday.

I spent two very cold nights looking for rodents there, and photographed one beautiful Maya mouse. In addition, the tiny forest was home to numerous rare orchids, birds, and insects. Next morning I found the owner of the land, and asked why he was cutting the last forest for miles around.

The owner, Señor Juan, was an old gentleman, very calm and polite. He wasn't happy about cutting the trees at all. But his land was poor, and he couldn't let any potential pasture go unused. Besides, he was going to sell the timber for about $30. He knew very well that oakwood could fetch fifty times as much in any port city, but transporting it out of the mountains would be extremely difficult. Perhaps, he suggested, I'd like to buy this patch of land? He'd jump at any opportunity to save the grove.

I was between two summer jobs and on a very tight budget. I had been sleeping in my car since I had left Boulder three months earlier. But Señor Juan agreed to sell me the land for $50. I couldn't resist.

It took me two days to drive a hundred miles over very bad roads to the nearest town. Two more days to get all the paperwork for the land sale and to buy some barbed wire (I needed to protect the forest from cows and goats). Then two days to get back. We built a fence, and I made a few signs: NO ENTRY—LAND MINES! I was probably the world's poorest owner of a private nature reserve.

I went back to the town and spent a day in local restaurants, telling some people that I was an American journalist investigating a KGB-run biological weapons testing site, and others that I was a Russian journalist investigating a location where the CIA had tested a "dirty bomb." I hoped it would be enough to make people stay away from the grove. Then I had to leave.

Señor Juan had little contact with the outside world, but his sons would travel to the town once every few months. Three years after buying the grove, I received a letter from him, informing me that a major storm had caused serious damage. I couldn't go there right away, but now, a year later, I had such an opportunity.

There was now a good highway running all the way across Guatemala, so I got to that town in just one day. A local bus delivered me to within forty miles of my land. I rented a horse and rode straight to my forest. What I saw was heartbreaking. The great oak had fallen. As it rolled down the steep slope, it broke a few smaller trees. Señor Juan was a very honest man: he didn't touch the wood, even though it was now on his land, and he fixed the fence. But he couldn't save the precious leaf litter, and it was all gone. As the shade had disappeared, tall grass took over the forest floor.

I thanked Señor Juan for everything he had done, and told him it was OK to sell the fallen oak. I spent two nights in the grove, but instead of the many species of rodents that used to live there, I found only harvest mice (a grassland species) and two flying squirrels living in a hollow in one of the broken trees. A pair of rare owls also remained. It had been raining for a few days, so I could safely burn some grass around the grove to create a firebreak. With all that tall grass inside, fire was a major danger. The grove was still home to some rare flora and fauna, but not to the species I tried to save.

As far as I know, nobody has seen a live Maya mouse since I found one in my reserve back in 2003, so this species is probably extinct. My desperate attempt came too late: the forest was too small to be a secure refuge.

Nowadays biologists get news of plants and animals going extinct almost daily. For every such species that we know about, there are probably dozens that disappear unnoticed. But you can never get used to this news. Not only because you know that every such loss is forever, that evolution never repeats itself exactly, but also because every species is a unique, wonderful miracle of nature, beautiful in its own way.

As you can imagine, I wasn't in a very good mood when I crossed the Mexican border and hitched a ride to Tuxtla Gutiérrez, the capital of Chiapas. There I met with Jeronimo Laso, who, despite being very young, was already one of Mexico's top croc experts. He was extremely knowledgeable; I think he knew every crocodile population in southeastern Mexico.

Jeronimo told me of an easily accessible place with Morelet's crocodiles—a group of small lakes straddling the Mexico-Guatemala border in the area known as Selva Lacandona (Lacandon Jungle) after the local branch of the Maya people. It was only half an hour's hike from a paved road to the first lake, so I got there in less than two days.

It was the least interesting time of the year to be in the forest: the dry season, when many frogs, salamanders, snails, and insects were hiding. But there were plenty of larger creatures around. Lakeshores were covered with animal tracks. I didn't know exactly where the international border was, but I guessed that my tent was set almost directly on it. Whenever I walked deeper into Guatemala, I saw animals that can be found only in places with little or no hunting: spider monkeys, guans (large forest wildfowl), and peacock-like ocellated turkeys. Agoutis and peccaries would sometimes walk by my tent in broad daylight. The lakes were full of fish and had clear water, so I didn't have to go back to civilization for supplies; all I needed to do was make a bow and a few arrows (iridescent turkey feathers, easy to spot on the forest floor, came in handy). The water was so transparent that I could snorkel safely, knowing that no crocodile would be able to sneak up on me (Morelet's crocodiles can kill people, but such cases are extremely rare).

The first croc I found was a dead juvenile floating in the water. I didn't know why it had died. I decided to use it to make a "jungle guest book," and hung it from a tree fifteen feet above ground, then raked the soil below to see what kinds of animals would leave their footprints. I chose a place where I could watch the dead crocodile while observing live ones in one of the lakes. The next morning there were jaguar tracks in my guest book. I waited, and soon photographed a young male jaguar as he sniffed the air, trying to figure out how to get the bait. As soon as he left, I lowered the bait to ten feet, hoping to photograph him jumping up if he came back. But instead, a rarely seen badger-like creature called a grison showed up the next morning, and later a couple of beautiful, otter-like cats called jaguarundis. Then I left the guest book unattended for a few hours . . . and found it hacked by the time I got back. The tracks showed that the jaguar had returned, pulled the dead crocodile down, and carried it off.

There were a few adult Morelet's crocodiles in each lake, but they didn't make any sounds. The only crocodiles "singing" were American crocodiles that lived in two large oxbow lakes a few miles farther into Guatemala. I was excited to find that they made more roars than those living in coastal habitats. It was a perfect confirmation for my theory—but there were only three "singing" animals, so the sample size was too small.

Why didn't Morelet's crocodiles "sing"? All literature and all researchers I had talked with claimed that these two species had their mating seasons at the same time. Was I doing something wrong? I soon found out.

Although my camp was in the forest away from the road, I saw people almost daily. They followed the path along the chain of lakes and passed through without seeing me or my tent. They were migrants from Central America, entering Mexico illegally to find work or to continue all the way to the US—not an easy journey, considering the number of police and military checkpoints on Mexican roads. Then one evening I returned to my tent and saw three armed men standing in front of it. They asked me who I was and what I was doing there. It was a tricky situation because I wasn't sure which country I was in, so I asked them first to tell me who they were. They said they were Zapatistas.

It wasn't good.

Just a few decades ago, Selva Lacandona was the largest remaining forest in Mexico and Central America. Local Lacandon Mayas lived in small communities deep in the jungle and practiced sustainable agriculture. Then people (mostly other Mayas) from outside the forest began moving in, clearing the land, and building numerous villages and towns. The government started a resettlement program, moving farmers from the dry north of the country to the south. What the bureaucrats didn't realize was that the parched cornfields of the north could actually feed more people than the poor rain forest soils of the south.

After the land was mostly cleared and the Lacandon Mayas were pushed into their three last villages, the government and various non-government organizations (NGOs) tried to save the remaining forest by limiting settlement, outlawing logging, and creating the huge Montes Azules Biosphere Reserve. The resulting tension was one of the main reasons for the massive popular uprising in the 1990s, led by a leftist group called the Zapatista Army of National Liberation (EZLN) after Emiliano Zapata, a Mexican revolutionary of the early twentieth century, who today would probably be called a Maoist.

For various environmental organizations, the emergence of the Zapatistas created a painful conflict of interest. Historically, most of them claimed to be protectors of indigenous rights as well as the environment, ignoring the fact that these two goals are conflicting more often than not, especially in today's developing countries, where most native peoples have become way too numerous to keep living off the land in traditional ways. Now these organizations had to choose between the people and the forest, because the EZLN demanded unlimited rights to logging, mining, and land privatizing. It called itself "an indigenous group," although it represented the recent settlers rather than the few remaining Lacandon Mayas.

Over the years, the bitterness of the conflict has subsided somewhat. The Zapatistas have mostly switched to nonviolent methods and are more visible online than on the ground. They achieved considerable autonomy for some parts of Chiapas, while the government and the NGOs have been able to convince a few communities to relocate out of the biosphere reserve. But deforestation continues, and some wildlife, such as tapirs and jaguars, is almost extinct in Chiapas. In addition, local people have become deeply suspicious (or paranoid, depending on your point of view) of scientific research in the area and cry biopiracy (patenting natural resources such as medicinal herbs without paying off the local "owners") every time someone tries to collect a fruit fly from the forest—ironic considering that those same locals often do their best to destroy those natural resources completely.

My talk with the three Zapatistas wasn't an easy one. They didn't like Americans, they were suspicious of biologists, and they hated people camping in what they considered *their* forest. I explained what I was studying, but it only made things worse. Their chief said that I was making it up. Everybody knew that crocodiles couldn't "sing," he said. He stared at me like a ferocious Aztec warrior preparing to sacrifice a captive to the war god. At that unnerving moment the youngest of them, who looked like a teenager playing paintball, suddenly spoke for the first time.

"Excuse me, Comandante," he said, "but they do 'sing.' I hear them sometimes when I hunt. They roar like a chain saw when you try to start it."

That instantly changed the situation. People living in remote places are invariably fascinated when you tell them something they don't already know about their environment. I had to give an improvised lecture on crocodilian mating habits. Soon it became obvious that Jorge, the young kid, also had a lot to tell. And one of the first things he told me was that American and Morelet's crocodiles had different mating seasons. All books and professional zoologists had been wrong. The way Jorge put it, American crocodiles mostly "sing" in February, when mornings are cool. As soon as mornings turn warm in March—"Just a few days from now," he said—they fall silent and Morelet's crocodiles start "singing."

I asked if he could stay with me for a few days (just to make sure I didn't commit any biopiracy or sorcery), and after a bit of persuasion his comandante agreed to give him a few days off.

Jorge was actually twenty-two years old. His mother was a Lacandon Maya, his father a Yucatec Maya from the Yucatán Peninsula. When the kid was ten, his father left to work in the US and was never heard from again. Jorge was too young to work the family field, so he became a hunter's apprentice and soon got so good at hunting that he managed to feed his mother and two sisters. Later the family moved to a larger village, and he signed up with the Zapatistas to do border patrols. This way he could still spend as much time as he could in the forest.

He stayed with me for five days. Now I was able to get data from two lakes at once. The timing of this unexpected help was perfect, because the next day I accidentally hit a snag while wading through a swamp and tore off my big toenail. Walking through the forest between the lakes was now taking a lot of time. Jorge brought a bag of rice, which was a nice addition to my diet of fish and nuts. And, of course, we had a lot of things to talk about: habits and idiosyncrasies of the animals around us, fine points of the art of tracking, conservation issues, Mexican politics—and sometimes girls.

In four days the weather changed, mornings became warm, and Morelet's crocodiles began to "sing." American crocodiles didn't shut up immediately, but gradually fell quiet after another week.

Just as in other crocodile species, only the largest Morelet's in each lake would ever "sing." I knew that at least one of those large animals was a male because I saw him make love to one of the smaller ones. Crocodiles virtually always mate in the water, the female almost submerged, the male wrapped around her and hugging her with all his feet. In this case the male was almost ten feet long, close to the maximum recorded size for Morelet's crocodiles, and the female less than five feet long, but they seemed to have no difficulty whatsoever, probably because the foreplay lasted more than half an hour.

All "songs" I observed were the same: infrasound and a short, loud roar. They sounded very different from the quiet "coughs" of American crocodiles but looked almost the same on spectrograms, graphic images of sounds that I later got from my tape recordings. I still don't know how these two species manage to remain separate, but perhaps hybridization is limited by a combination of two mechanisms: the differences in "songs" and in the timing of the mating seasons.

I didn't see any headslaps, but later I found that Mexican biologists had observed Morelet's crocodiles headslapping in captivity. Perhaps I didn't watch them long enough to see that. In any case, they roar way more often than headslap, just like my theory predicted for a species living in small lakes and forest swamps.

Every time I meet people like Jorge, I feel sorry that I can't magically transport them to some good university to be trained as biologists. I stayed in the forest for another week after he left, but after spending a full month there I decided it was time to move on. I walked back to the road, made a quick side trip to Bonampak, the largest ancient Maya city in Selva Lacandona, with some impressive ruins, then crossed Guatemala and half of Honduras, and took a ferry to Útila.

Utila is one of about a dozen small islands off the Caribbean coast that were once parts of British Honduras (its mainland portion is now Belize). Nowadays these islands are divided between Honduras and Nicaragua. Many local residents are descendants of British pirates, but their main source of income today is dive tourism. These islands are the cheapest place in the Western Hemisphere to get all kinds of diving certificates.

Sarit and Alex, my candidates for the African expedition, were working in a small dive center, training for dive master certifications. I stayed with them for a week, diving on coral reefs, exploring bat caves, watching my toenail grow back (it would take a full year to complete the process), and generally having fun. Dmitry and Olga, a couple of my Russian blog readers who were traveling in

Central America, came from the mainland to meet me and get an autograph. I talked them into learning to dive. After the first dive Olga surfaced crying; she said she didn't speak enough English to understand her instructor. Under the stress she didn't realize that he was using only signs underwater. But soon they overcame all fears and got the coveted plastic cards.

I got my own advanced diving certificate on Útila, and after the standard amount of transport-related adventures caught my flight home from Guatemala.

Children of illegal settlers, Chiapas

19
Island Romance
Crocodylus rhombifer

A male that invents a novel way to impress a female
puts himself above all competition.
—Konrad Lorenz

THE ONLY CROCODILE SPECIES IN THE NEW WORLD that I hadn't observed yet was the Cuban crocodile. I could easily arrange a trip to Cuba, but I couldn't find reliable information on the timing of the mating season. Some sources claimed it was March, while others said May. And I couldn't spend a long time there, because I'd miss the alligator mating season in the States.

I decided to observe captive Cuban crocodiles, although I knew it would be a poor substitute for observations in the wild. There was a breeding pair in Zoo Miami, so I contacted the zoo's curator of reptiles, who looked through his records and found that the pair mated in late March or early April. He gave me a free pass to the zoo, which I really appreciated. Now all I had to do was go to the zoo at six o'clock every morning and spend five hours a day in front of the crocs' enclosure.

The Cuban crocodile is the most terrestrial of living crocodiles. It has muscular feet with non-webbed toes, and is very good at walking, running, and even jumping. It's very intelligent and can hunt in packs.

There's a famous BBC video showing these crocs leaping their full length out of the water in an attempt to snatch a hutia (a large arboreal rodent) from a tree. Currently hutias are the only large mammals in the freshwater swamps of western Cuba where these crocodiles live, so the crocs feed mostly on fish and turtles. However, prior to the arrival of humans on the island about four thousand years ago, Cuba had at least eight species of ground sloths. It's possible that Cuban crocodiles have evolved their terrestrial habits to chase those sloths. Some fossil sloth bones show marks left by crocodile teeth.

The Cuban is also one of the rarest crocodilians, with only a few thousand left in two large freshwater swamps. In the past it was more common, also occurring in the Cayman Islands, the Bahamas, and possibly Florida. It can hybridize with the American crocodile, which is also present in Cuba. Recently it was discovered that the American crocodiles of Florida still carry some Cubans' genes from past hybridization. When Cuban crocodiles became rare, hybridization apparently became more frequent, and is considered one of the main threats to the species' existence.

Cuban crocodiles grow to about ten feet. Despite their relatively small size they are said to be the most dangerous crocodilians to keep in captivity because they are very smart, fast, and aggressive. They have particularly massive teeth in the back of their jaws, used for crushing bones and turtle shells, and can tear off a human hand with just one rapid jerk of the head. One zookeeper told me that they reminded him of the pack-hunting "velociraptors" (actually dromaeosaurs) in *Jurassic Park.* They are usually dominant over larger American crocodiles (who, in turn, are dominant over even less aggressive American alligators). But experienced animal trainers have taught them to behave in a friendly fashion and to respond to individual names as well as various voice commands. There was a circus animal trainer in Germany who kept a seven-foot Cuban crocodile as her house pet. The croc lived in the bathtub and walked freely around the house, but never attacked any visitors (not even children) or the house cat. In summer she would walk it outside on a leash.

These crocodiles have an unusual "marbled" color pattern, with small irregular black spots covering the entire body. They look very beautiful compared with their dull-colored relatives. The pair I was observing had been living in Zoo Miami for a long time, but they were still curious about people. Every time I showed up near the glass fence of their enclosure at six in the morning, the male would approach me to investigate. But he didn't react that way to the crowds of visitors who arrived later in the day, except for an occasional glance toward some particularly plump toddler.

The male roared daily, but I saw a headslap only once. Later I observed captive Cuban crocodiles in other zoos, and all of them roared rather than headslapped, just as my theory predicted for a species living in shallow swamps and small lakes. This was getting boring: getting information on each new species took a lot of effort, and my predictions were correct every time. But the sample size was growing painfully slowly. Still, I was determined to get information on as many species as I could. By now I had at least some data on about half of them.

Once the male roared, the female would usually approach him looking for love, and that's when my observations got interesting. The couple did something that nobody had ever seen crocodiles do. The female, which was much smaller than the male, would climb on his back, and he'd give her a ride around the pool, making two or three circles. Was it something all Cuban crocodiles do during courtship, or something this pair had invented? I don't know yet, but I hope to find out someday. The rest of the foreplay was more typical: a lot of tender chin-touching and nose-rubbing, eventually leading to gentle hugging and then to a passionate embrace.

In case you are curious, crocodilians make love the same way we do, although the male has to evert his penis, which is otherwise hidden inside the cloaca (the single multipurpose opening that crocodiles have under their tails). Females have a tiny penis-like clitoris, also hidden inside. Males of other extant reptiles such as snakes and lizards have two penises (although they use only one at a time), but ostriches and a few other relatively primitive birds have one, similar in structure to the crocodilian penis. Since dinosaurs' position on the evolutionary tree is between crocodilians (their close relatives) and birds (their descendants), we can assume that male dinosaurs also had one evertable penis.

This method of inferring the unknown features of extinct groups by looking at their surviving relatives and descendants is called phylogenetic bracketing. One other thing we can infer by this method is that, since both crocodilians and cassowaries (a kind of ancient-looking ostrich living in New Guinea and northeastern Australia) use infrasound for signaling, dinosaurs probably used it as well. One is tempted to think that the ultra-long necks of colossal sauropod dinosaurs were used as resonators for producing really loud infrasound, probably allowing them to communicate over hundreds of miles.

In 1994 crocodilian penises suddenly made headlines and even got discussed by the US Congress. It was found that male alligators living in heavily polluted lakes near Orlando had abnormally small penises. Chemical pollutants were disrupting their hormonal functions. The increased public awareness soon led to similar discoveries in fish and mammals, including humans, but many effects of environmental pollution on our own chemistry remain unknown.

The pair at Zoo Miami stopped all mating-related behavior after just one week. I thought it happened so soon because the female got pregnant, but that year, for the first time in about a decade, she didn't build a nest or lay eggs. Maybe they were getting too old, although they weren't particularly big for Cubans.

I had some time left before the beginning of alligator mating season, so I decided to go to Texas and Louisiana and look for possible study sites. It was also a nice excuse to drive through Tennessee and finally meet Nastia in person after almost a year of being friends online. I had been planning this date for quite some time.

We decided that I'd get to her place just as she would come home from work on Friday. Of course, as soon as I was a few hundred miles away from home, my old Toyota burned its transmission. I had to drop it at a shop and rent another car, a huge, brand-new Chrysler, certainly not something I'd ever choose to drive. I was yet to learn that Nastia wasn't the kind of girl to be impressed with such cheap glamour. I drove the rest of the way at speeds I personally considered safe and rode up to Nastia's apartment block five minutes before the appointed time.

I already knew that she was very smart (one of the best Ukrainian math students of her generation, according to the results of various competitions), talented (it was her writing style that first got me interested), adventurous (she was twenty-one when she left Ukraine for a country where she didn't know anyone), and had a great sense of humor. But what did she look like?

She opened the door, and I saw that my long drive was worth it. Her brilliant writings were very deep, mature, and often in a minor key, so I expected to see someone broody and reserved. Instead I met a sunny blonde, soft-spoken but sharp-tongued, with beautiful golden-green eyes and a charming smile. I instantly nicknamed her "Little Mermaid."

The next day we went to the Great Smoky Mountains for a short hike. My mother, who is always very compassionate about my personal life, called me from Moscow to warn me not to flip logs in the forest in search of salamanders. "Don't let the poor girl know how crazy you are," she said. But I knew that if I didn't manage to get Nastia interested in zoology one way or another, our chances for a serious relationship would be close to zero.

I love Appalachian spring even more than the famous fall colors. The forest was all illuminated with blooming redbuds and dogwoods. In some places there were so many flowers on the ground that trees seemed to be standing in blue lakes. And, of course, there were salamanders under every log: slimy salamanders with their night sky–like pattern, tiny dwarf salamanders, bright-red efts, and countless others.

Looking back now, I can say that I've succeeded. Nastia has become a really good nature photographer with a particular interest in amphibians and other small animals.

Some people are born naturalists. I got the gene from my father, who is an organic chemist but also an expert in Central Asian butterflies. He managed to survive through perestroika by doubling as a butterfly hunter for Western collectors. My parents got divorced just before I was born, and I didn't meet him until I was twelve, by which time I was already obsessed with biology. I think I was a pretty clean nature-versus-nurture experiment, even if the sample size was small.

But the world of field biology is so fascinating that almost everyone can be pulled in given a chance. I did my best to seduce Nastia to its charms.

I stayed in Tennessee for two weeks instead of two days but then had to leave to get my car back and continue with my alligator studies. I managed to persuade Nastia to visit me in Miami before flying to Europe for much of the summer.

I was so happy that she turned out to be so wonderful that I couldn't sleep much and covered a lot of ground in ten days of driving. I found two really interesting study sites along the Gulf Coast, decided to look for the third one later, and drove home.

I've never put as much effort into cleaning my apartment as I did before Nastia's visit. I felt certain that it wasn't just another date; she was the person I wanted to remain in my life. I even evicted all spiders, though I didn't touch the geckos—I considered them the main attraction of the place.

When Nastia arrived, she wasn't particularly impressed, and I decided to cut our time in that apartment to a minimum. So I gave her the usual tour of the Everglades, introduced her to my study subjects, and drove her to Key West. I was going to take her to the best place for a romantic date in the eastern half of the country.

West of the Florida Keys there's a group of tiny sandy islands called the Dry Tortugas. There's no drinking water there and no permanent population, just seabird colonies and a nineteenth-century fort. The huge hexagonal fort covers the entire island it's on, except for a small beach. Four of its six sides jut into the sea, protected from the surf by a brick wave-breaker that forms a ring of shallow lagoons around the fort. The fort itself is very beautiful, and the views from its walls are stunning, with turquoise shallows stretching to the empty horizon in all directions. There's plenty of wildlife: migratory birds stop by during their Gulf crossings; nurse sharks and lobsters appear in the lagoons at night; the world's largest barracudas patrol the surrounding shallows. But there are no biting insects.

The islands are a national park, so there are a few rangers working in shifts at the fort. Most people come there on day trips, but it's much better to camp on the beach. You can snorkel under the coral-crusted brick walls, walk around the fort along the moat, or explore the endless dark passages inside. And the best thing to do is to grab your sleeping bag and climb on top of the fort late at night to watch the stars. No other place east of the Rockies has so little light pollution, and the Milky Way is so bright you can see its reflection in the sea. It's a great place to visit with a loved one. But you should never come there alone, because not having someone to hug on such a romantic island would be intolerable.

Three days later I was on my way to alligator swamps, and Nastia was flying across the Atlantic. She had made all arrangements for her trip to Europe before we met, and couldn't change them. I had to leave for Africa two weeks before her return. We wouldn't see each other for seven months. But I knew that I wanted to be with Nastia, no matter how long it would take.

20
Scientific Testing
Alligator mississippiensis

There ain't no surer way to find out whether you like people or hate them than to travel with them.
—Mark Twain

I WAS SLEEPING IN MY KAYAK IN THE MIDDLE of a quiet blackwater bayou, surrounded by tall bald cypress trees. The warm night air was full of birdsongs, blinking fireflies, and the sweet aroma of young leaves, and swamp mosquitoes hadn't emerged yet.

Around midnight I woke up as the boat rocked slightly. I expected to see an alligator or a beaver swimming by, but it was a mermaid. She rose to her waist out of the water and was looking at me, her elbows on the side of the kayak. Her tender skin seemed very white in bright moonlight; she had long blond hair and golden-green eyes. I moved away a bit, and she slid into the boat. Instead of a fishtail she had a pair of very slender legs. If not for her magic eyes—the color of a spring morning—one could easily mistake her for an ordinary girl, just an unusually beautiful one. I put my arm around her shoulders, still cold and wet, and she smiled, warming up in my hands.

When the stars began to grow faint, she smiled again, kissed me good-bye, and disappeared in the water without a splash. I fell asleep in my kayak, which suddenly felt empty and uncomfortable.

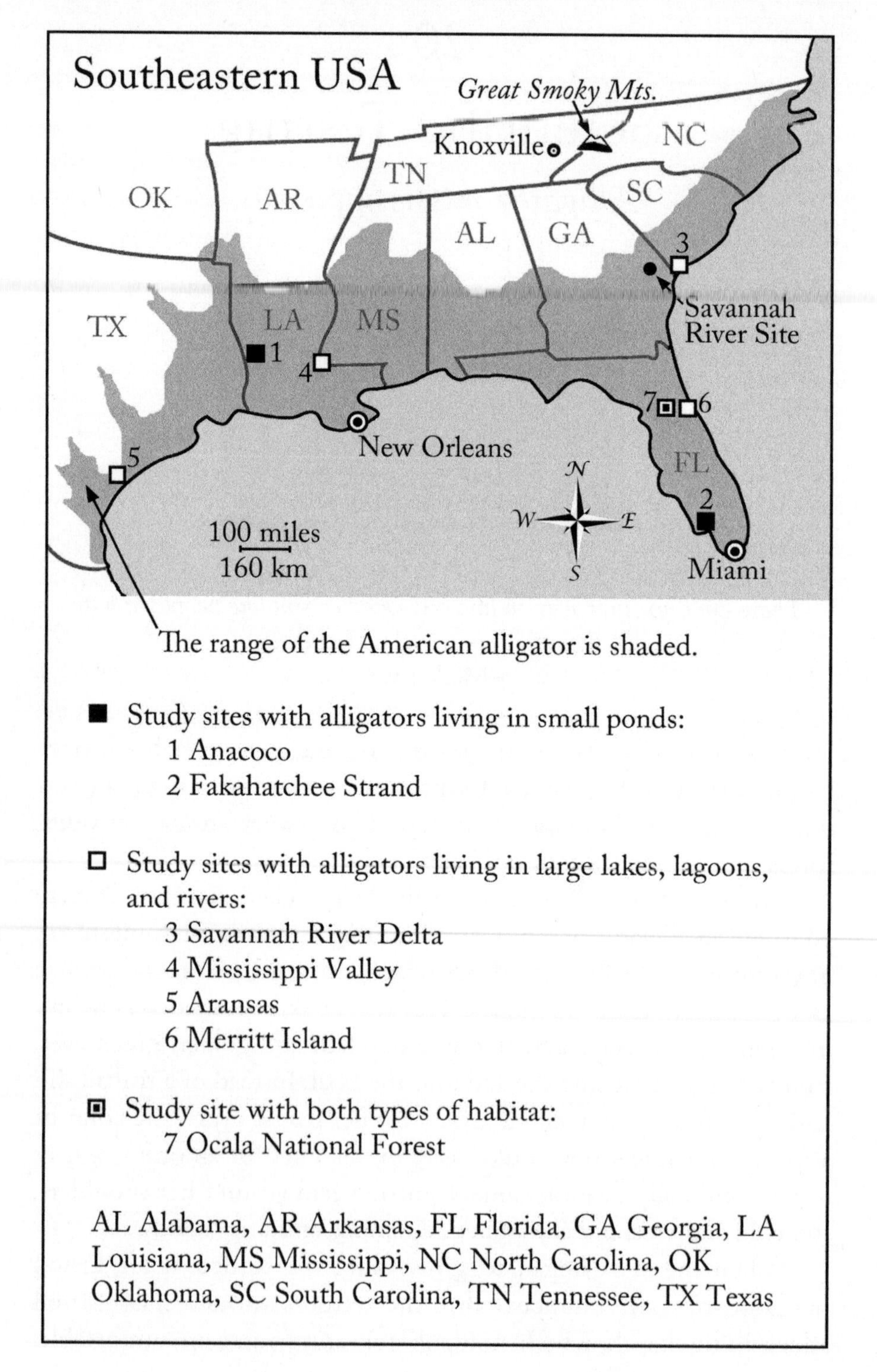
Southeastern USA
Great Smoky Mts.
Knoxville
NC
TN
OK
AR
SC
AL
GA
3
Savannah River Site
TX
LA
MS
1
4
7
6
New Orleans
FL
5
N
W
E
S
2
100 miles
160 km
Miami
The range of the American alligator is shaded.
■ Study sites with alligators living in small ponds:
1 Anacoco
2 Fakahatchee Strand
□ Study sites with alligators living in large lakes, lagoons, and rivers:
3 Savannah River Delta
4 Mississippi Valley
5 Aransas
6 Merritt Island
▣ Study site with both types of habitat:
7 Ocala National Forest
AL Alabama, AR Arkansas, FL Florida, GA Georgia, LA Louisiana, MS Mississippi, NC North Carolina, OK Oklahoma, SC South Carolina, TN Tennessee, TX Texas

Soon the first alligator bellow woke me up again, a powerful growl that made the water, the boat, and even the trees on the shores tremble.

The mermaid would come every night, as soon as the shining moon rose above the tops of bald cypress trees. But the moon was rising later and later, until one night the thin crescent didn't show up until dawn, and my green-eyed guest didn't come at all.

I deflated the kayak, found my car, hidden in the bushes near the road's end, and drove away, but every slender young willow at the roadside reminded me of those magic golden-green eyes in the silence of the night.

This time the kayak rolled heavily, and I woke up for real. A black bear was drinking from the nearby shore, puffing and sniffing. The air smelled of dawn, but there were still stars in the sky. A few alligators floated silently around, never moving a muscle. I was in Louisiana, and it was my third week in the swamps.

I began my alligator observations in Texas, in Aransas National Wildlife Refuge. It's an extensive area of coastal lagoons, reedbeds, and salt marshes. Every winter lots of people come there on boat tours to see the only completely wild whooping cranes left in the world. By late April the cranes leave, but bird-watchers stay for a few more weeks, hoping for the so-called *fallouts*, sudden spikes in the numbers of migratory birds that happen when strong northerly winds prevent them from moving and they get stranded in woodland patches.

Adult alligators in and around Aransas lived in large lakes and lagoons, so I had to paddle for hours to find them, while packs of javelinas (peccaries) were stealing food from my camp hidden in an oak grove. By early May, the heat and humidity became unbearable, and clouds of horseflies emerged. When I recorded all fifty alligator "songs" I needed, I was really happy to leave Aransas and move to the shady forests of Louisiana, where the mosquitoes were also happy when I arrived.

I needed to find two types of study sites: places where adult alligators lived only in large lagoons, lakes, and rivers but not in small ones; and places where they all inhabited small ponds. The first kind was easier to find; there're plenty of places like that along the east coast, Aransas being one of them. But I didn't want them all to look identical, so I found one such area farther inland, on the Louisiana-Mississippi state line. It was a stretch of Mississippi Valley that was completely flooded at that time of year, so it was all one huge body of water. Alligators didn't like the cold, fast Mississippi, so they were hiding in flooded forests at the edges of the floodplain. The water there was up to ten feet deep.

As the murky waters of the Mississippi filtered through those forests, they became transparent, and snorkeling there was no less interesting than in the Amazon. The Southeastern US has the most diverse freshwater fauna outside the tropics. Only the rivers of East China come close. (Sadly, much of Chinese freshwater fauna is already extinct, and in the US hundreds of freshwater species are endangered.) Sometimes I saw five or six turtle species in one hour, including exquisitely beautiful map turtles and giant alligator snappers. Humongous catfish lurked in deeper pools; Jurassic-looking paddlefish patrolled the channels. Once I met an even more impressive "living fossil," an armor-plated alligator gar the size of my kayak. I didn't have to worry too much about safety. This area was close to the northern limit of the range of the American alligator, so most gators were too small to be dangerous. Although even alligators six feet long have been known to kill people, usually they have to grow to ten or twelve feet to have the self-confidence for hunting adult humans.

Unfortunately, being at the northern edge of their range also meant that their population density was low. Finding enough adult males in dense woods was almost impossible. I realized that I wouldn't be able to obtain the sample size I needed in one season, so I decided to take a break and look for another study site.

This time I needed a place where alligators would live only in small ponds. That proved tricky, because almost everywhere in the South there was a dense network of rivers. Even a single river running near my study site would "disqualify" it.

I found only one good area. It didn't have an official name, but I saw it signed "Anacoco Floodplain" on old maps. There were a few small rivers there, but they were half-dry (unlike the flooding

Mississippi, they didn't get their water from the snowmelt hundreds of miles to the north) and heavily overgrown. My previous experiments had shown that underwater sounds don't carry well along such streams. Besides, I didn't see any alligators in these rivers; they were all hiding in "gator holes" deep in the woods. It was the dry season, the easiest time of year to walk around. Still, finding those holes was very time-consuming. Fortunately, I got help.

Ten years earlier, I happened to live in California during the time of the "Internet bubble," when anyone capable of writing his name in English could get a job as a software tester. I had some Web designing experience, so I got hired as a senior Web designer for a small start-up in San Francisco. The company was owned and financed by a rich guy named Will, and my supervisor was Sandeep Singh, a young, always-smiling British Indian. It went well for a while, but when the "bubble" began to burst, Will stopped paying our salaries. Sandeep and I ended up working for a few months for free, and there was nothing we could do because Will had hired us as outside contractors and our contracts were very vague. I lost only a few thousand dollars, but Sandeep had bought a lot of equipment on his own, and he was devastated. He had serious debts, which was a big issue for an honor-bound Sikh. As if this wasn't bad enough, his wife, quiet and charming Sumita, was diagnosed with cancer. Sandeep had to give up his lawsuit against Will and return to Britain to get her medical care. She died the following year.

Sandeep's hobby was studying the culture (or, rather, cultures, because there are many) of African Americans, particularly those still living in the isolated, deeply traditional towns of the South Carolina islands and the Gulf Coast. After Sumita's death, it took him five years to repay all debts, but now he was back on his feet, and he traveled to the Deep South every year. When I heard he was in New Orleans, I invited him to help me with alligators.

Anacoco lies on the northern edge of the Cajun lands. It is as "Southern" as it gets; in one remote forest hamlet people told us that we were the first whites to set foot there since the slavery era. Of course, it was still the US: roads were mostly paved, any place of significance had a decent library, and you could sometimes drive for hours never seeing a plastic bottle or a piece of paper on the roadside.

Working in the swamps was difficult for Sandeep, who was definitely not an outdoorsman, but he took it stoically. With his help I managed to get all the data I needed by early June.

The result was the same as in other study sites across the South. My theory predicted that alligators living in large bodies of water would bellow less often and headslap more often than those living in small ponds. It was only half true. They headslapped a lot less in small ponds but didn't bellow more. The frequency of bellowing was the same everywhere. And I had no idea how to explain this difference.

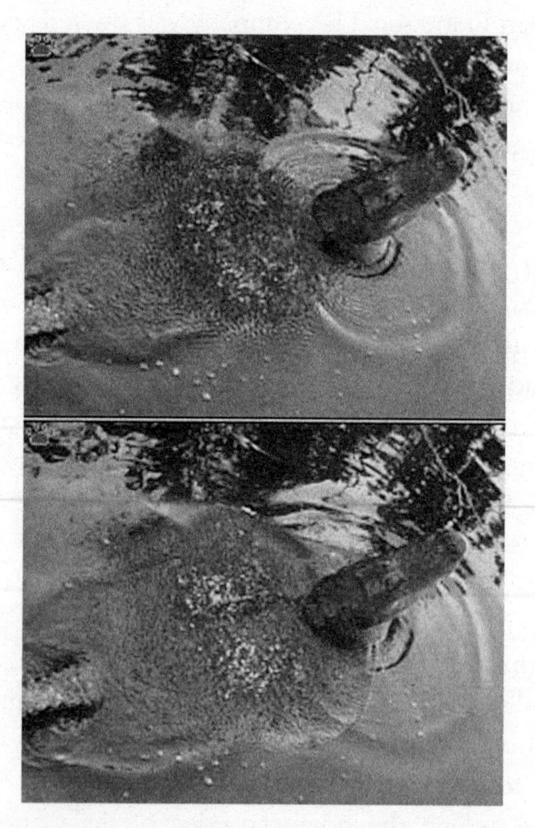

Patterns of "water dance" above the lungs of an infrasound-producing alligator

★ ★ ★

I drove back to Miami, met with Sarit and Alex, and we spent a couple days running around town getting ready for the African trip. Suddenly, there was a lot to take care of: my driver's license was about to expire and there was an unpaid speeding ticket from years before that suddenly appeared; Sarit's passport had no empty pages left; I needed to pack all my stuff; Alex had to fix their "hippie bus" (an antique Volkswagen minivan), which they had used to drive around Mexico and Central America the previous year. Finally, everything was ready, I rented a truck, hooked up my Toyota, and we drove to a patch of savanna near Orlando where a friend of mine had just bought a homestead. I left all my books and furniture in his shed, Alex parked the "hippie bus" behind the house, and we returned the truck then drove north in my little Toyota, packed so tightly that it took Sarit two minutes to get out whenever we stopped.

We went to North Carolina to dive with sand tiger sharks at a World War II German submarine wreck, then to Brooklyn where Sarit's mother lived. I flew to Denver, met with my mother who arrived from Moscow, rented another car, and drove with her around the West, from the Grand Canyon to Yellowstone, for three weeks.

When I returned to New York, we called Shirley, our fourth team member. It was time to go to Canada for two weeks for a compatibility test. Everything had been discussed before, but suddenly Shirley began changing plans. She would have to leave two days later, she said, because of a dentist appointment. OK, we understood. The next day she said that she would have to come back three days earlier. That was already a bit of a stretch, but we agreed to shorten the trip. Then she said she had some other business that had come up suddenly, so she would fly to Canada and meet us there for just two days.

That was too much. She had four months to get ready, and still she was completely unprepared. If she couldn't arrange to leave home for a few days, what would it be like in Africa? So I told her that I couldn't take her there. I felt awful about this, but in hindsight it was the right decision. The trip to Africa would be more difficult than anyone had anticipated. Having an inexperienced person with us would make it the disaster it almost became anyway.

The three of us drove north along the coast, from New York to Quebec. Nowhere in North America can you get the kind of problems you have to deal with in Africa, but I was determined to make this test as real as possible. I had to get to know my companions well while it was still possible to part ways. Will they start whining after three days of almost nonstop driving? Will they complain when it gets really cold, or really dangerous, or when there's no food? I'd take them on a whale-watching trip every time there was a storm. Soon the weather got so cold and wet that we'd sleep in the car rather than pitch the tent; Sarit had to crawl halfway into the trunk so we could all fit in. We went diving wherever we could, first in wetsuits, then in dry suits. In a small town on the Gulf of Saint Lawrence, we went looking for giant Greenland sharks. The water was almost freezing, our suits were leaking, and the visibility was so poor that we almost bumped into twenty-foot-long sharks before seeing them. Some of the sharks couldn't see us either, because parasitic crustaceans were covering their eyes, but we knew that they could feel the electric fields of our bodies from at least a few feet away. We hoped they'd be able to tell that we weren't seals, their common food.

After two weeks of this, I had to admit that I had perfect companions. They never complained and always seemed reasonably happy. We drove back, dropped my car at my relative's place in New Jersey, hauled our luggage to JFK, and finally got sixteen hours of rest in the air before arriving in Johannesburg.

An alligator "dance" (nighttime courtship gathering). On rare occasions, alligator dances continue until dawn.

Alligator couple on its way to the dance

American alligators in love

Baby American alligator

Huanlong Lakes, Sichuan

Chinese alligators, Anhui Province, China

Sensory pits—visible in the photo as small dots—on the snout of a baby American alligator

Yacare caimans in Bolivia. Yacare caimans often carry water hyacinths on their backs.

The infamous "Death Road," Bolivia

A mugger crocodile (left) and an Indian gharial (top), Kataraniaghat Wildlife Sanctuary. Muggers have snout shape typical for omnivorous crocodilians, while gharials are specialized fish-eaters.

Female muggers trying to seduce a large male in response to his impressive roar

A well-camouflaged mugger trying to attract birds looking for nest material, Madras Crocodile Bank, India

A tigress trying to take me off a tree, Kanha National Park, India

A tick and a fly on the tigress's face

An American crocodile (left) and an American alligator. Crocs have triangular snouts, and both upper and lower teeth are visible when the mouth is closed. Gators have shovel-shaped snouts, and only upper teeth are visible.

American crocodiles kissing, Dominican Republic

An American alligator in a roadside ambush, Merritt Island, Florida

A spoonbill sandpiper chick in Chukotka, Russia. Fewer than two hundred breeding pairs remain in the world.

Adult male gharials. The structure at the end of the male's snout is called a *ghara* in India, meaning "cooking pot."

A gharial couple, female on top. Kataraniaghat Wildlife Sanctuary

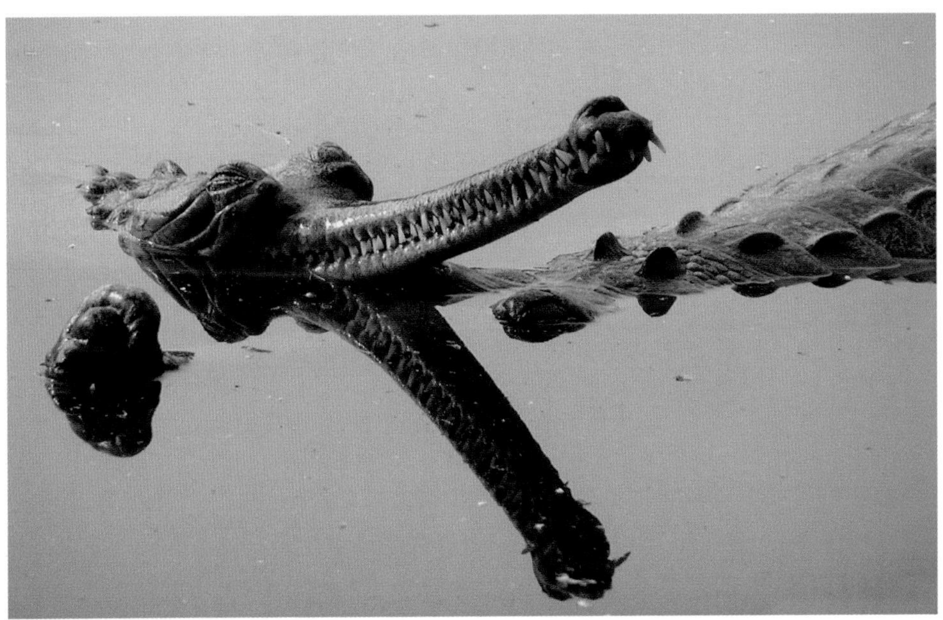

Amazon riverboat

Black caiman,
Mamirauà Reserve

Sleeping arrangements on an
Amazon riverboat, Brazil

Roaring yacare caimans in the Pantanal, Brazil

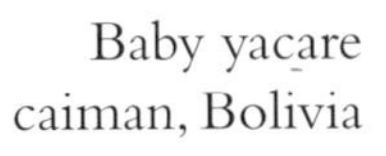

Baby yacare caiman, Bolivia

Victoria regia, a flower and a young leaf, Guyana

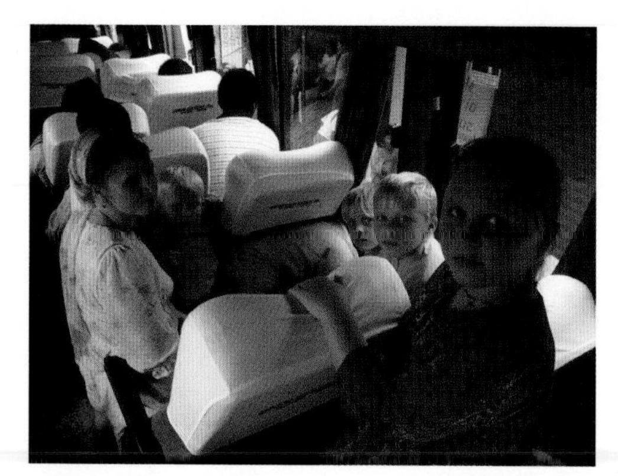

Russian Old Believers in Bolivia

A sloth crossing a flooded river valley, Bolivia

Pampas deer fawn, Bolivia

Orinoco crocodile, Venezuela

Close-up of an Orinoco crocodile. This species was saved from extinction by John Thorbjarnarson and his colleagues.

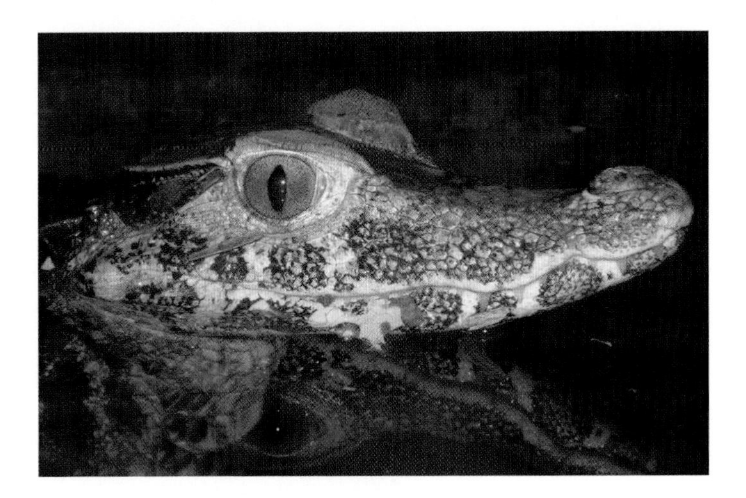

Schneider's dwarf caiman, Venezuela

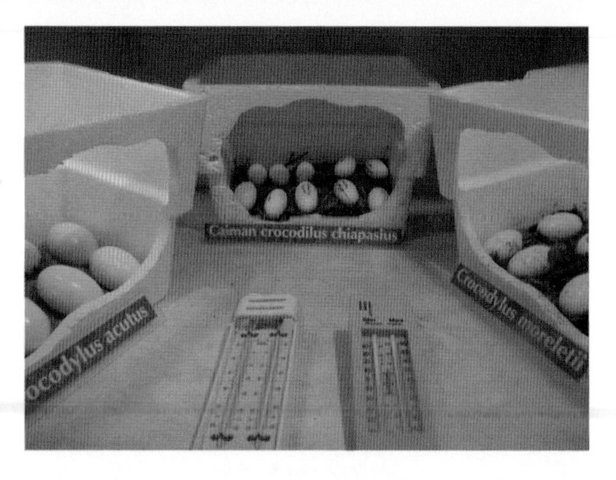

Eggs of spectacled caimans and two species of crocodiles in a breeding facility, Chiapas, Mexico

Morelet's crocodile, Rio Chajul Valley, Chiapas

The fort at Dry Tortugas, Florida

Lioness chasing a springbock, Etosha National Park, Namibia

A watering hole in Etosha

Columbus in Namaqua Desert, South Africa. This was our ride through eight African countries for four months.

Palmated gecko, Namib Desert, Namibia

Namaqua chameleon, Namibia

Puff adder, near Pretoria, South Africa

King protea, Table Mountain, South Africa

Baby Nile crocodile

Nile crocodiles, Namibia

(Left) A curious leopard contemplating leaping into our jeep, Saint Lucia, South Africa

(Right) Lioness bloodied from eating a buffalo, South Luangwa National Park, Zambia

Comet moth, Ranomafana National Park, Madagascar

Mouse lemur, Kirindi Nature Reserve, Madagascar

A dominant Nile male, Saint Lucia Lagoon, South Africa

Our team in Afar, Ethiopia

A Christian boy,
Tigrai Province,
Ethiopia

A typical view of an Afar village

Our obligatory guides at Dallol Volcano, Afar

Erta Ale lava lake, Afar

Lava lake exploding

Narrow-snouted crocodile, Gabon

Central African dwarf crocodile, Gabon

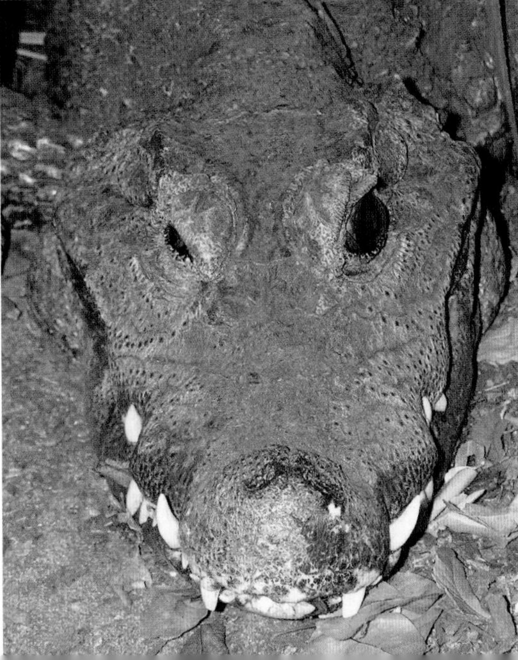

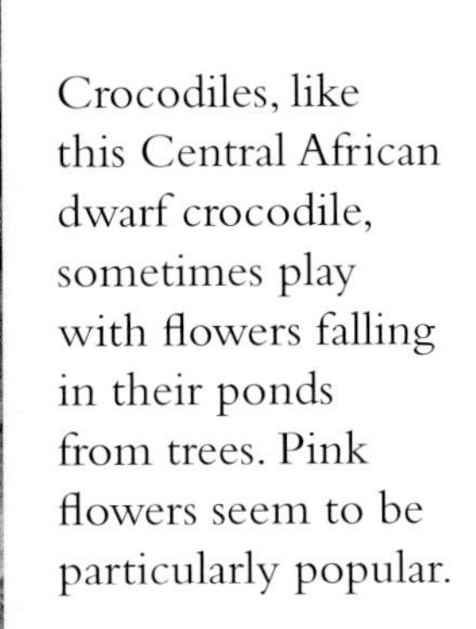

Crocodiles, like this Central African dwarf crocodile, sometimes play with flowers falling in their ponds from trees. Pink flowers seem to be particularly popular.

White-mustached guenon, Korup National Park, Cameroon

Baby chimp orphaned by bushmeat trade, Cameroon

Great white egret with a baby alligator, Florida

Tomistoma, aka false gharial, St. Augustine Alligator Farm

Spectral tarsier, Tangkoko-Batuangas National Park, Sulavesi

Nastia, scuba diving in Komodo National Park, Indonesia

Giant ant, Kerinci National Park, Sumatra

A traditional effigy of a dead relative, Tana Toraja, Sulawesi, Indonesia

Mimic octopus mimicking a sea anemone, Lembeh Strait, Indonesia

A saltwater crocodile enjoying a soft pillow on a hot day, Raja Ampat Islands

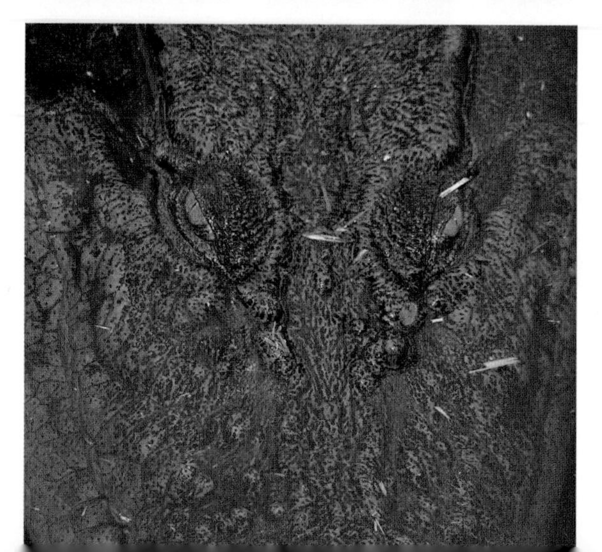

Saltwater crocodiles are very good at stalking prey from underwater. Sorong, Indonesian New Guinea

Siamese crocodile, Madras Crocodile Bank, India

Old broad-snouted caiman, Itaipu, Brazil

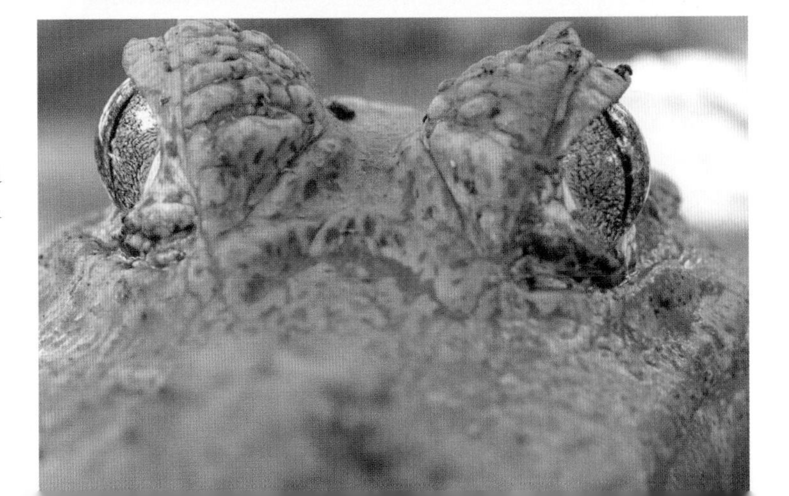

Broad-snouted caiman

Baby broad-snouted caiman, Bolivia

Forest lake in Yasuni National Park, Ecuador

As for national parks, getting all research permits would be totally impossible. I didn't know in advance where I would find the conditions I needed for work, so I'd have to obtain permits for dozens of parks in many countries. That would cost a fortune and take years. We came up with an easy way to solve this problem. We decided to write a guidebook, the first guide to Africa in Russian, focused mostly on nature. A large travel company that I often wrote articles for was interested in publishing the book. Now we had journalist business cards and could apply for park permits as travel writers.

That worked beautifully. In most countries we easily arranged meetings with local officials, up to the level of deputy minister, and obtained blanket permits to visit all parks for free. I was very careful to always mention my crocodile research, but nobody cared about my little "side job" as long as we offered free promotion. We were determined to return those favors as well as we could.

We landed in Johannesburg on a freezing winter midnight, rented a tiny Fiat, and drove to Cape Town almost nonstop. South Africa is a big country; even though roads are excellent, crossing it takes days. When we arrived at the office of Jackal Buyback, we were told that the truck wasn't ready: steering problems had to be fixed.

We ended up waiting for it for two weeks, but we didn't mind. If I had to choose a city to be stuck in for two weeks, I'd probably choose Cape Town anyway. It's located in a unique part of the world called the Cape Region. Local flora is so diverse and unusual that botanists consider this relatively small area to be one of five or six major biogeographical domains, equal in importance to North America, Europe, and nontropical Asia combined. Nobody knows why it's so rich in plants. One possibility is that its California-like climate hasn't changed for a particularly long time. One type of shrubland called *fynbos* ("fine bush" in Afrikaans) has hundreds of endemic plant species, such as proteas with flowers the size of a dinner plate. You don't even have to leave town to see lots of wildlife: Nile geese nest on building roofs, baboons raid gardens, penguins breed on rocky shores, whales mate within a stone's throw from piers,

21
Navigating Africa
Crocodylus niloticus

It is a capital mistake to theorize before one has data.
—Arthur Conan Doyle

GENERALLY POOR AND UNDERDEVELOPED, Africa is nonetheless an expensive continent to travel in. Most people go there to see wildlife, but it is sparse outside national parks and reserves, which aren't cheap to visit—some charge exorbitant fees. And you need your own transportation to enter most of them.

I realized that the study I had planned in Africa would be impossible unless I could cut expenses. Just renting a four-wheel-drive truck fo three months would cost about $10,000. Fortunately, a year earlie Hyena Overland, a large Cape Town–based tour company, created subsidiary called Jackal Buyback for leasing trucks. They were using business model that is very popular in Australia, called "buy-back": y buy a truck for a slightly elevated price, and, after using it for a f months, you sell it back to the company for that price minus the cos the lease. We arranged in advance to lease a truck from them.

When my mother heard about the buyback scheme, said: "That's not good. You'll never get your money back." Jackal seemed to be a respectable company, and online reviews excellent.

and hippos live in a small nature reserve. If you try scuba diving, you can play with inquisitive fur seals and small friendly sharks.

We quickly decided that our book would include information not only about national parks, but also about private tour operators. Immediately we were offered free whale-watching tours, white-shark diving, and walks with a baboon patrol (a group of rangers that follows bands of baboons around, preventing them from raiding gardens and breaking into houses). Being journalists was more fun than we expected.

We felt a bit uneasy about that sudden bonanza. What if we couldn't deliver? Doing any kind of business in Russia is extremely risky. An average life expectancy of a company is less than a year, and the book market is poorly developed and totally unpredictable. Just to ensure that our consciences remained clean, we decided not to ask for any trips organized specially for us: we'd go only if there were other customers, so that our presence wouldn't cost the operator extra money.

We were curious to see how the country was dealing with the legacy of the apartheid, which had ended just fifteen years earlier. We expected to find it deeply divided along racial lines, and extremely sensitive to any color-related subjects, the way the US was before the Obama election.

But we were wrong. We talked to people of all backgrounds, from white business owners to unemployed residents of the Cape Flats, a giant slum near Cape Town (we spent some time there trying to photograph dune mole rats). Everybody discussed racial issues freely, everybody was deeply concerned about crime, brain drain, and other problems—but there was no racism, black or white, and no overblown worries about political correctness. The only really xenophobic speeches we ever heard were by local blacks who didn't like the influx of Zimbabwean and Mozambican immigrants. We were yet to learn that in some neighboring countries the situation was very different.

Finally, our truck was ready. Jackal Buyback had a huge office, half garage, half repair shop. All their trucks had personal names; we got "Columbus." We loaded it with camping gear, fuel and water

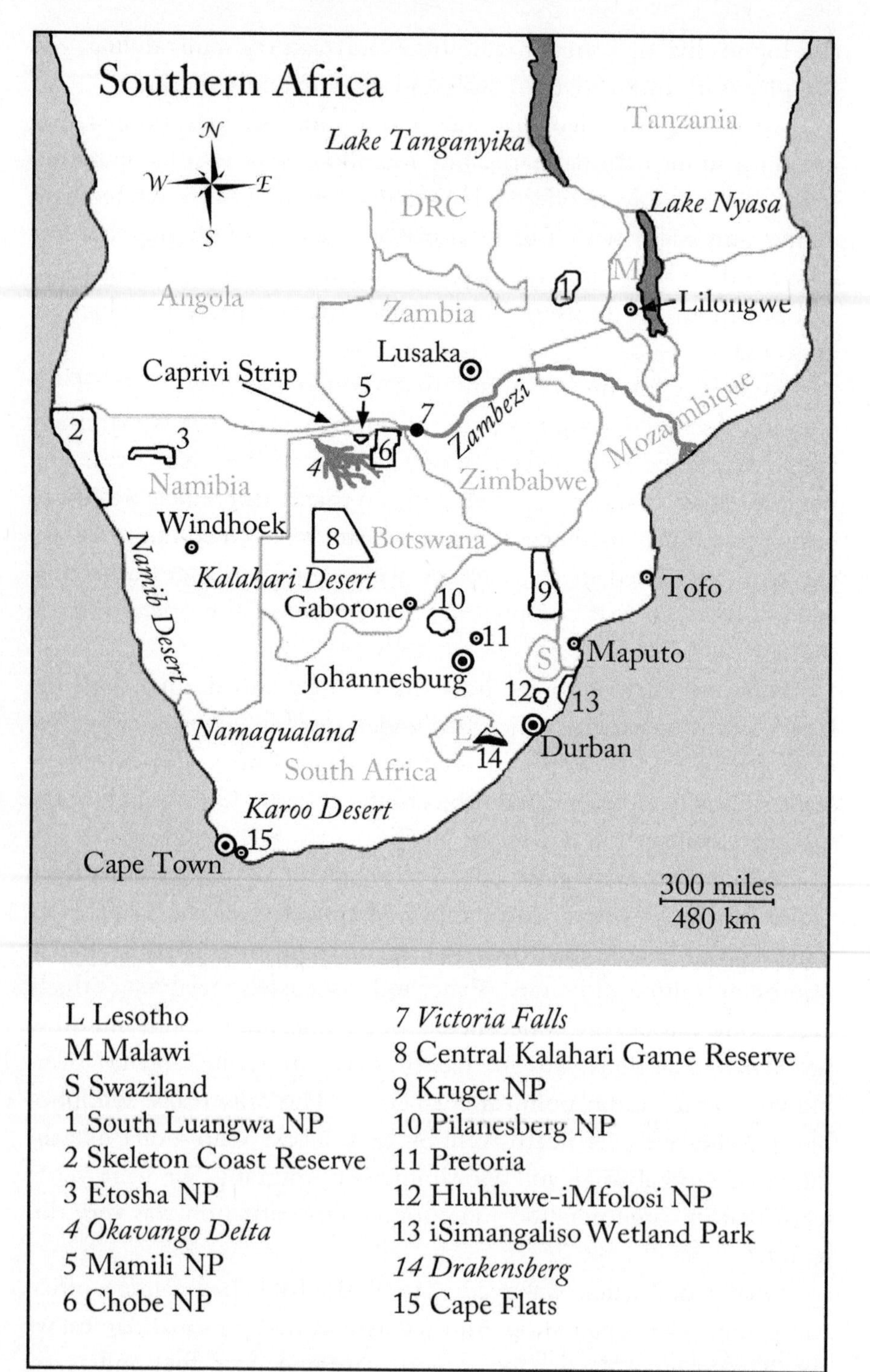
Southern Africa
N
W
E
S
Lake Tanganyika
Tanzania
DRC
Lake Nyasa
M
Angola
Zambia
Lilongwe
Lusaka
Caprivi Strip
Zambezi
Mozambique
Namibia
Zimbabwe
Windhoek
Botswana
Namib Desert
Kalahari Desert
Tofo
Gaborone
Maputo
S
Johannesburg
Namaqualand
L
Durban
South Africa
Karoo Desert
Cape Town
300 miles
480 km
L Lesotho
M Malawi
S Swaziland
1 South Luangwa NP
2 Skeleton Coast Reserve
3 Etosha NP
4 Okavango Delta
5 Mamili NP
6 Chobe NP
7 Victoria Falls
8 Central Kalahari Game Reserve
9 Kruger NP
10 Pilanesberg NP
11 Pretoria
12 Hluhluwe-iMfolosi NP
13 iSimangaliso Wetland Park
14 Drakensberg
15 Cape Flats

canisters, spare tires, and all other stuff needed to survive in the bush. And we were on our way.

As soon as we got on a freeway, we realized that the steering hadn't been fixed. The wheel play was so huge that we almost crashed into upcoming traffic a few times. By the time we got to Namaqualand, Columbus was practically impossible to drive. We had it repaired and hoped that was the last mechanical problem. It wasn't. Something would break almost every day.

But the country we were traveling through was so stunningly beautiful that we just didn't care. It was early August, the first month of South African spring, and the desert was blooming. I'd seen splendid desert blooms in Asia and the Americas, but Namaqualand was on a totally different level. The main flower here was a bright-orange daisy; fields of these daisies looked surprisingly similar to the California poppies of the Mojave Desert. But there were also scarlet, yellow, and lilac daisies, dozens of varieties of ice plant, blooming aloe—we could count a few dozen species every time we stepped out of the truck. Unbroken flower beds stretched for hundreds of miles. Springbok gazelles, ostriches, mountain zebras, and blue cranes seemed to be floating in a sea of orange paint. It was hard to believe that within a month this country would be dry and lifeless.

Then the blooming plains ended, and the desert began to look like the Africa of children's books: granite mountains, dry canyons, umbrella-like acacias, herds of antelopes and giraffes. Telegraph poles were decorated with giant haystacks—communal nests of thousands of little birds called social weavers. Sun-bleached grass reminded me of Nastia's hair. I couldn't bear the thought that she was missing all these wonders.

We still had a few weeks before the crocodiles' mating season, so we stopped at every park or reserve to gather data for our guidebook. Many animals were already breeding: we watched baby hyraxes, giraffes, zebras, elands, and bontebok antelopes. As you can imagine, these park visits were the kind of work we totally hated and did only out of duty. Our route was a broad zigzag, from the cold seashores, where we looked for gannet colonies and rare dolphins, to the red dunes of the Kalahari, where we met cheetahs, warthogs, and lions. Needless to say, Columbus didn't miss a chance to get stuck in the middle of a lion pride. We found that its alternator belt was too long and had fallen off. All national parks had beautiful campgrounds with power outlets, hot tubs, and kitchens. Hot tubs were particularly welcome, because nights were still freezing.

The local people were neither black nor white. They were Khoisans, the oldest race on Earth: delicately built, with yellow skin and almost Oriental faces. Their languages were full of strange clicking sounds. Their ancestors had created some of the world's most beautiful rock art, hidden in remote mountains all over Southern Africa. Unfortunately, their cultures had been almost totally destroyed, and most of their settlements looked much worse than the poorest Indian reservation in the US. The nomadic Bushmen of the Kalahari, once the only people known to live in an area with no surface water, belonged to that race. But they had all been forced off their lands.

We entered Namibia. The country looked almost uninhabited, with gravel roads cutting through endless deserts, game ranches, and nature reserves. In the early twentieth century, German colonists killed much of the population, except in the far north of the country, and seized almost all the land. Later, the so-called "veterinary cordon fence" was erected, cutting off the densely populated North, officially to prevent the spread of cattle disease but in reality to create a huge reservation for the blacks. The fence wouldn't be removed until 2010.

We were slowly moving along the edge of the Namib Desert, until one evening Columbus broke down in the middle of nowhere. Its transmission fell apart. Alex and I had to walk for miles to find one of the countless luxurious safari lodges scattered in the area. The owners, an old German couple, were very nice to us. They towed Columbus to their lodge and offered to take it to Windhoek, the capital. They were going there in a big truck the next day, so they could easily put Columbus in the trunk. We were unbelievably lucky.

I rode in Columbus with their maid, a woman from an Ovambo tribe, while Sarit and Alex got in the big truck cabin with the lodge owners. The drive took five hours, and for all that time these nice folks went on an endless tirade about "lazy Negroes destroying the country." They also said that they were very happy to have us with them. "Otherwise," they said, "we'd have to take our maid in the cabin, and ... you know ... people would be talking. We businessmen aren't supposed to mix with lower races." It all felt like the 1930s.

It took us two days to get Jackal Buyback to pay for the new transmission. It wasn't our last visit to a mechanic. Small things kept breaking every day; we started sending weekly reports to Jackal, signed, "Still alive—Vladimir, Sarit, and Alex."

Windhoek is a pleasant city by African standards, with the world's best rock shop (the owner has discovered a few new minerals himself) and relatively compact slums. Within a day we managed to get permits for unlimited stays in all national parks, for night driving, and even for unguided walking—the holy grail of African travelers.

In most African parks you aren't allowed to leave your car, except in designated fenced campgrounds and picnic spots. In South Africa, even having your elbow out of the car window is a violation. In some places you can also go to unfenced "bush camps," but that costs a lot of money. All other ways of avoiding those safety regulations are also very expensive, so the whole thing looks a bit like an extortion scheme. Night driving is usually not allowed, despite the fact that most animals are nocturnal. Tourists seldom get to see much except for sleeping lions and grazing ungulates. In the morning, when campground gates are unlocked, there's often a traffic jam as everybody tries to get out and see something interesting before it gets too hot for the animals to move. Picnic sites aren't fenced and attract a lot of wildlife. If a lion wanted to grab a tourist, a picnic site would be the best place. Do all these draconian measures improve safety? I doubt it. In the few remaining places where people can walk freely, such as in Mana Pools National Park in Zimbabwe, the rate of animal attacks is actually lower.

Anyway, in Namibia and later in some other countries we managed to avoid all these restrictions. We could see the famous red dunes of Sossusvlei at sunrise (it's an hour's drive from a gate that is only unlocked at dawn); we could explore the Skeleton Coast without paying $750 per day for staying in a lodge; and later we could study crocodiles at night and at dawn, when they were the most active. I'd never get such permits for research, but being a journalist opened all doors.

Namibia is the best country in Africa to visit for a short vacation. It's relatively safe, the roads are reasonably good, there's no malaria, the wildlife is no less abundant than in much more popular Kenya and Tanzania, and park fees are relatively low. The entire coast is desert, but it's one of the most interesting deserts in the world. Farther inland are rocky mountains, savannas, and, in the far north, dry forests and extensive seasonal wetlands.

I particularly enjoyed our long drives through roadless deserts, where we sometimes saw no sign of human presence on the planet for days. Not all deserts could be driven across: in some areas we had to be extremely careful to avoid damaging the thin, almost invisible lichen crust that would take centuries to regrow. In such places we had to use roads made of rock salt.

There were few plants and animals in the desert, but the ones we could find were worth looking for. Once we drove by a giant *Welwitschia* plant. This "living fossil" looked like a tree stump the size of Columbus, with two fifty-foot-long leaves spreading in curls around it. There were foxes, geckos, black desert chameleons, gemsbok antelopes, and brown hyenas. But the most interesting animals were darkling beetles. Their methods of surviving in the desert were amazing. Some had complicated systems of channels inside their backs; at night, the beetle would stand head-down on a dune crest, letting the fog condense on its body, and the channels would siphon water droplets toward its mouth. Others could run so fast that the wind would cool them down in midday heat. Yet others learned to communicate by sound, tapping on the sand with their bellies.

After driving through the desert for another week, we turned inland, into the dry savanna with little villages scattered through the hills, herds of cattle with humongous horns kicking up dust on the road, and chains of women carrying buckets of water on their heads from distant wells. Days got really hot, and nights were finally warm. Every evening we'd find a narrow side road, follow it into the woods, choose some opening between shrubs, and camp. That's called "bush camping." It used to be the normal way of spending nights while traveling almost anywhere in Africa, but few people do it anymore, mostly because an average tourist today is too paranoid to sleep without being locked into something. I usually tried to sneak out of the tent for at least an hour every day, either in the morning or in the evening, to give Sarit and Alex some time alone. This habit let me see a lot of beautiful sunrises and enjoy some interesting night walks.

Once we were invited to stay in a Himba village. The Himbas are an unusual tribe that has migrated to Namibia from East Africa centuries ago. Once a year they organize an expedition to Angola to mine red ochre, and then cover themselves with ocher mixed with animal fat from head to toe every morning. They don't wear much else, except for leather miniskirts and lots of ankle bracelets. Ankles are the only body part you should never let others see, they believe. We stayed in a hut with a girl of fifteen. She had had a lover since she was thirteen, but she was going to get married soon: in Himba culture marriage has little to do with sex. Being a desert tribe, the Himbas use smoke, rather than water, to clean themselves every morning; they squat over a pile of smoldering herbs and let the smoke get into every pore. The method apparently works well, because they look very clean and healthy. Our hostess was so beautiful that Sarit fell in love with her. Unfortunately, it didn't go anywhere because we couldn't stay there for long.

Our last stop before getting to crocodile-inhabited areas was Etosha National Park. If I ever have children, this will be one of the first places I'll take them. The park is almost the size of Massachusetts, but during the dry season most of its wildlife concentrates near a few watering holes, the best of which sits at the edge of a tourist camp, with benches on two sides and a low fence separating people from the pool. An endless procession of animals marches in from the savanna to drink. In twenty-four hours, you can sometimes see fifty thousand antelopes, zebras, giraffes, elephants, rhinos, warthogs, lions, leopards, cheetahs, hyenas, and jackals. They all behave differently. Giraffes, for example, are very shy and would sometimes wait for hours to have their turn, while young elephants are ill-mannered delinquents and would hang around the pond for an extra half hour just to keep the giraffes from drinking.

Etosha is a favorite of wildlife photographers. I noticed that they had a certain hierarchy, based on the size of one's camera lens. Sarit always carried a long lens, and if one of those macho photographers noticed that her lens was longer he got visibly upset. I had only a tiny camera, but if they tried to make fun of me, I told them that, first, I didn't need to compensate for a lack elsewhere, and second, the better a photographer you were, the closer you could get to your target. That drove them mad sometimes. Still, I liked them a lot more than trophy hunters.

We couldn't force ourselves to leave Etosha for a whole week. Once we spent a day watching a lone lioness. She was hiding near a small watering hole in a patch of knee-high grass the size of a dinner table. You could walk to within twenty feet of her without realizing she was there. Hundreds of zebras, wildebeest, and springbok gazelles were watching the grass patch from three hundred yards away. Most of them already knew she was there, but new animals kept arriving, all of them thirsty. They could sense the danger, so they'd inch toward the pool for hours before getting close—and then the lioness would charge. In twelve hours we saw her try four times, but she always charged a minute too early or a second too late, and the wave of panicked game stampeded away without losses. You'd think watching this for a whole day would be boring, but we were totally exhausted from the emotional strain by the time we left.

Our first crocodile study site was the Caprivi Strip, a narrow band of Namibian territory bordered by Angola, Zambia, and Botswana. Like the rest of the country, it has a dry climate, but rivers flowing from elsewhere turn it into an extensive seasonal wetland, part of the immense complex of floodplains, swamps, and inland deltas stretching across much of Zambia and northern Botswana into Zimbabwe. Caprivi has three wonderful national parks, but they get few visitors: most tourists go to the much-promoted Okavango and Chobe in Botswana. In Mamili National Park we were the only people except for six park rangers. We were absolutely free to explore the flooded savanna, dense forests, and papyrus swamps. Being on our own had its downside, though. One morning Columbus got stuck in the middle of a swamp, and we spent a whole day digging it out. I don't know if we'd have succeeded; the rangers' truck happened to show up, and they pulled us out.

There were lots of crocodiles. We found one area where they all lived in a large river with steep banks, not in small ponds or swamps—a perfect site for my study. The mating season was just beginning. It had been five weeks since we arrived in Africa, and I was eager to get to work. We set up camp. Next morning I got out of the tent two hours before sunrise and walked along the riverbank, listening. The problem was, I didn't really know what to listen for.

The Nile crocodile was the first crocodilian species to become known to Western science. Herodotus, a Greek historian who lived in the fifth century BC, left an excellent description of its appearance and natural history. He observed crocodiles in Egypt, where they were kept as sacred animals, the representatives of Sobek, the god of the Nile River. But Herodotus didn't say anything about their mating rituals.

Later, numerous scientists have studied Nile crocodiles. Part of what they have written is nonsense. For example, a few highly respected naturalists have described a small bird that cleans crocodile teeth. Such a bird apparently doesn't exist, although once I photographed a school of tiny fish cleaning the teeth of an American crocodile. But some of these researchers were very accurate in most areas and observant, and spent decades studying crocs. And their accounts contradicted each other. A few of them described loud roars produced by Nile crocodiles. Others went into great detail about this species' signaling but didn't say a word about roars, mentioning only headslaps. Something strange was going on.

The sun was slowly rising, dark purple because of the heavy haze. Although much of the Caprivi Strip was flooded, the water had come from places far to the north. Around Caprivi it was the height of the dry season, with lots of wildfires.

Hippos were calling from deep pools hidden in papyrus swamps; birds were singing everywhere. A large crocodile was floating silently near the shore, probably touching the bottom with its feet. It raised its head and tail, the water on its back danced for a brief instant as infrasound was produced—then the crocodile headslapped. I was only thirty feet away, so I could hear a soft "coughing" sound just before the slap. I waited for three more hours, but nothing else happened.

I thought I had the answer: Nile crocodile "songs" didn't include loud roars. I was yet to find out that things were much more complicated.

22

Masters of the Bush

Crocodylus niloticus

Everything in Africa bites, but the safari bug is worst of all.
—Brian Jackman

The Nile crocodile is an amazing animal. No other reptile can adapt to such diverse habitats. Before the invention of firearms, it occurred almost everywhere in Africa, from rain forests and mangroves to deserts, from papyrus swamps to the African Great Lakes, from the far south of the continent to Palestine. On Madagascar, Nile crocodiles live in caves—the only crocodilians known to do so. In the dry mountains of the central Sahara, small Nile crocs once lived in tiny pools hidden in deep canyons. They were the surviving relics from the times when the Sahara was wet and green. Alas, in the twentieth century French colonial officers shot all those "canyon crocodiles" for target practice.

The Nile crocodile has been studied on and off for twenty-five centuries, but one of the most unexpected discoveries was made only a few years ago. It was found that the Nile crocodile is not one but two species, almost identical in appearance, but different genetically. One is slightly smaller and less aggressive than the other. Currently, only the larger species (still called the Nile crocodile) occurs in the Nile River, but a few centuries ago both species lived there. All crocodile mummies from Ancient Egypt belong to the smaller species. Apparently, the priests of the god Sobek knew the difference and bred only the smaller, less dangerous kind in their sacred pools. So the scientists who rediscovered the smaller species named it "sacred crocodile," and chose the word once used by Herodotus for its scientific name, *Crocodylus suchus*.

The ranges of the two species are still poorly known, but it appears that "true" Nile crocodiles today live mostly in East and Southern Africa and on Madagascar, while sacred crocodiles occur in the western part of the continent. That new data hadn't been published when I was in Africa, so I didn't even try to study sacred crocodiles. Later I had to look for them in American zoos. All my studies in the wild were on Nile crocodiles. As far as I could find out, both species are habitat generalists, and their "songs" are very similar.

Molecular studies also showed that the sacred crocodile was the ancestor of all four species of crocodiles that live in the New World. It was a sacred crocodile female that managed to swim across the Atlantic. The Nile crocodile is a much younger species.

There was too much work to think about crocodile systematics when we were in the Caprivi Strip. We had to get up well before dawn, and that was a good thing. Days were hot and boring; early mornings and evenings were pleasant and busy. We quietly sneaked out of the tent and walked along the river, trying to avoid getting caught between a grazing hippo and the water. All around us, the night shift was going to sleep and diurnal animals were waking up. Frogs and crickets fell silent one by one, while the bird chorus was gaining strength. We were all ears, trying not to say a word the whole morning: that was the time when crocodiles would "sing." We separated, reached places where large males lived, and waited for them to do something. They'd usually headslap once or twice every morning, with or without the "cough" sound.

A couple hours after the sunrise, crocodiles started crawling onshore to bask. They turned their tails sideways to catch the rays of sun with the huge scales (called *scutes*) running along the upside of the tails. These scales are rich in blood vessels and are used for rapid warming-up. They also make swimming more effective; for all crocodilians, the tail is the main source of propulsion in the water, while legs are used only for steering and keeping balance.

Most crocs opened their mouths and kept them open all day. Nobody knows why they do it; the theory is that it helps prevent brain overheating, but they also do it in cold weather, while alligators and caimans seldom bask with their mouths open. For us, seeing crocs leave the water was a signal that there would be no more "singing" that day.

By that hour all hippos were back in the river, and huge herds of buffaloes, lechwe antelopes, sable antelopes, and waterbucks were walking away from the meadows after their morning grazing, while zebras and impalas were moving in the opposite direction, from the woods to the meadows. We had already seen thousands of them, but we still couldn't help stopping in awe when a tiny zebra colt ran after its mother just ten feet from us, or a male kudu paused to look at us, or a golden puku antelope suddenly jumped out from the reeds. These creatures were too beautiful to get used to.

We ate our breakfast on the way to the camp. A pack of crackers, a mango, a small can of pineapple juice. There was never enough, but we didn't care.

Myriads of colorful birds were everywhere. On other tropical continents you mostly find birds in rain forests, but African savannas are very old, and their bird diversity is amazing. When we approached our camp, it always looked like there was a cloud of smoke over it. It wasn't smoke. A small puddle near the tent was the favorite drinking spot of queleas, a kind of social weaver. These sparrow-sized birds move around in immense flocks, sometimes up to half a million. You can hear the noise of their wings from a mile away.

We tried to spend the hottest hours in the river. There were dense reeds around, and a large crocodile wouldn't be able to sneak through them unnoticed. A general rule is that you should avoid reedbeds in Africa, because the water there can be infected by bilharzia, a very unpleasant parasite. But our river was too fast for that. The water was cool, sweet, and clear enough for snorkeling.

When the heat subsided, we left the camp and went looking for more crocs to observe the next morning. We tried not to get distracted too often. The savanna was teeming with life, from beetles to elephants, and if we didn't see something interesting for five minutes, there were always animal tracks to read—it was like walking across an endless newspaper page. Sometimes we had to stop and wait while a herd of elephants or a pride of lions crossed the road in front of us.

We returned to the tent late at night. Night walks were the best. We soon learned to identify most local animals by the brightness and color of their eyes in the beams of our flashlights. The brightest eyes belonged to relatively small lemur-like creatures called bush babies. Genets, wildcats, dormice, civets, bushbucks, bat-eared foxes, bushpigs, squirrel-tailed rock hares, aardwolves, springhares, small antelopes, skunk-like African polecats—all of them had slightly different eye colors. Once we found an animal that had no eyeshine at all. It was a pangolin, a termite-eating mammal that looks like a three-foot-long spruce cone. Pangolins are very difficult to find, so later we only had to mention having seen one to be taken seriously by park rangers, safari guides, and wildlife biologists.

In Africa you can instantly tell how well a traveler knows this land by asking what animals he is interested in seeing. First-time tourists are usually looking for the "Big Five"—the elephant, rhino, buffalo, lion, and leopard. These are, of course, fascinating animals, but there are thousands of other, no less amazing, creatures in the bush. Experienced naturalists cringe when they hear the words "Big Five." This term has been overused to death, just like the word "safari," or "aloha" in Hawaiian. Tell your guide that you'd like to see a pangolin, an aardvark, a caracal, or a suni, and he will respect you, even if his job will get more difficult.

We cooked our dinner, usually some pasta, canned fish, and tea. Then we downloaded photos to our laptop, put camera and flashlight batteries into our six chargers, and went to sleep. Alex had built a strange-looking contraption that allowed us to charge everything from the truck battery without turning on the engine.

The night was full of frogs' metallic ticking, nightjars' clicking, owls' hooting, spotted hyenas' "whooping," and lions' roaring. If I pressed my ear to the sleeping mat, I could often hear the low rumbling that elephants use to talk over great distances. They can hear it with their feet.

We usually had four or five hours to sleep, but we could hardly wait to get up.

When we were done with Caprivi crocodiles, we drove into Botswana, another sparsely populated country with an arid climate (the name of its currency means "rain" in the Tswana language) and extensive floodplains full of wildlife. It has quarantine fences, too, crisscrossing the entire country, running for hundreds of miles. They have nothing to do with apartheid. Botswana had been ruled by whites for only a short time, so interracial problems are virtually nonexistent. The fences were originally built to protect cattle-growing areas from wild animals and the diseases they carried, but now the population and its cattle herds have grown so much that the fences protect the wildlife from herders. Unfortunately, the fences interfere with zebra and wildebeest migrations, often causing the deaths of hundreds of thousands from thirst.

We entered Botswana from the far north. The capital, Gaborone, was in the far south, and we didn't want to drive there for two days just to get a permit. Driving in Africa is an interesting but tiring exercise. You can never relax for even a second, because a deep pothole, a suicidal child, or a galloping kudu antelope can show up in front of you at any moment. But we were told that the local authorities probably wouldn't be able to issue the permit we needed. They said we could observe the crocodiles without a permit, but we'd have to pay regular park fees. That's a lot of money. Botswana officially encourages only rich tourists to visit, and its parks are very expensive.

We had to wait for the officials to make the final decision, so we went to a cheaper park called Central Kalahari Game Reserve for a few days. The reason it's relatively cheap is that few people ever go there. It's the size of New Hampshire but has only two ranger stations and a few campsites. It's an endless sea of red dunes, with patches of low shrubs and a few trees along ancient river channels that have been dry for millennia. Only the toughest animals, such as gemsbok antelopes, brown hyenas, and honey badgers can survive there. Many of them have never seen people and are unusually tame. The roads are just tracks in the sand. There are campsite showers, but you have to bring the water with you.

On our third day in the reserve we saw a cloud of dust on the horizon. Another truck! We turned in that direction. When we got close, people in the truck shouted: "Have you seen the lion?" "What lion?" we asked. They were amazed. "You aren't looking for the lion? What are you doing here, then?"

They explained that the dry valley we were crossing was famous as the home range of the largest lion in the reserve. Black-maned lions of the Deep Kalahari are the largest and most beautiful in Africa, so we immediately decided to find that male. We soon saw his tracks, almost twice the average size. After a few hours of easy tracking over red sand we got to a bunch of small bushes with tracks going in and not out. But the lion was nowhere to be seen. We stopped and waited.

"I have to pee," said Sarit after a while.

"Be careful," we told her.

She opened the door, closed it again, and whispered, "Oops. . ." She had almost stepped on the lion's tail. The giant male was lying in the shade of our truck. I've seen hundreds of lions, but this was something really extraordinary, a massive old male with a black mane the size of an open umbrella.

The next night we were driving along another dry river valley and saw a porcupine. It was running away from us along the wheel rut. "I want to photograph it from the front," said Sarit. I knew how to do it with American porcupines, so I said: "Alex should get out, run around the porcupine and overtake it. Then the porcupine will turn its quill-covered back, which is better protected, toward Alex, and face us." Alex jumped out of the car, and we discovered that African porcupines were better runners than American ones. For a few seconds, Alex and the animal were running in front of our truck side by side. Neither of them could get ahead of the other. Then I had to stop: I didn't want to stress the porcupine too much.

When we had almost run out of fuel and water, we left the Kalahari and drove back to the civilized part of the country. The officials told us they couldn't issue the permit after all. We decided to leave

Botswana. It was easier for us to go to Zambia, the next country, than to drive to Gaborone and back.

As we were driving toward Zambia, we noticed a huge garbage dump on the outskirts of a small town. African garbage dumps can be good places to see wildlife. Sarit wanted to photograph marabou storks, so we walked inside. Suddenly, a large bull elephant appeared from behind a pile of garbage, chewing on an old newspaper. Local people who were sorting through stuff on top of that pile started shouting and throwing rocks at him. The elephant was visibly angry. We couldn't turn our backs on him, so we kept photographing. Then he charged.

Elephants can make mock charges and attack charges. In a mock charge, they try to look big and scary, so they spread their ears, raise their trunks, and trumpet. In an attack charge, they pull back the ears, coil up the trunk to make it ready for a strike, and move silently. The trunk is a lethal weapon; one swipe can break a cow's neck. This time it was a mock charge, so the elephant stopped about a hundred feet from us. He totally succeeded in looking big and scary. I grabbed Sarit's arm to make sure she wouldn't try to run away but found myself pulling her back from the elephant. She was way too eager to get a close-up photo.

We came back to that dump at night, looking for hyenas, and found no people and a large herd of elephants happily munching on scraps.

Elephant behavior differed remarkably depending on location. In places with little or no hunting, such as in Chobe and Kruger National Parks, they were mostly indifferent to people and would sometimes graze around our tent. In places with some native poaching, they didn't mind cars but could be aggressive or paranoid toward pedestrians. In areas with lots of hunting, they were always nervous and dangerous.

We saw only one attack charge the whole time. We were staying in a lodge in Zambia where we watched a large male walking aimlessly between bungalows. The lodge had no other visitors at the time, so the manager didn't mind. The elephant was close to us but didn't seem to care about our presence. I decided to take a photo of my friends with the elephant in the background, and turned my back to him for half a second. When I looked at him again, he was just twenty feet away, running straight into us. I made three quick steps toward him, raised my hands, and shouted something rude. He braked with all his four feet and ran away. I didn't get a single photo, but Alex obtained a nice shot literally from underneath the elephant's tusks.

Lions, mostly big males, would also do mock charges sometimes. They'd run at us, roaring, and stop so close that dirt from their feet would fly into our faces. They were a bit more predictable than elephants, so we did get a few good photos. Lionesses defending small cubs are known to do attack charges as well, but we were lucky never to experience one.

Our journey through northern Botswana, Zambia, and Malawi was a strange time in our lives. For decades, all three of us had been striving to excel in the art of travel on extremely tight budgets. So in Africa we mostly did the same, eating the cheapest food available, bargaining over everything (Sarit was incredibly good at this), and never sleeping under roofs or in paid campgrounds unless we absolutely had to. But once in a few days we'd get invited by a lodge manager eager to have his place mentioned in our guidebook, and we'd spend a night in decadent luxury. The more remote the lodge, the more exquisite and lavish it would usually be. Some places charged thousands of dollars per night and could be normally accessed only by chartered airplanes. The roads leading to them were so bad that we got used to spending, on average, two hours a day fixing Columbus. At first being able to stay in such places for free was fun, but we soon tired of it. We didn't really need this opulence. We didn't enjoy listening to people discussing the advantages of various private jets and the difficulties of finding well-schooled butlers. Bush camping was so much better. Eventually, we started avoiding lodges, stopping there only occasionally to use the Internet and talk to safari guides about local crocodiles.

Once in Zambia, we made a quick stop at Victoria Falls, on the border between Zambia and Zimbabwe. Unfortunately, it was the worst time to visit Zimbabwe. The country was rapidly disintegrating; its population couldn't get rid of Robert Mugabe and his army of thugs. All neighboring countries were full of Zimbabwean refugees, black and white. The inflation rate was so bad that you had to pay up front for restaurant dinners because the price would be different by the time you finished. One trillion Zim dollar bills were popular souvenirs on the Zambian side of the border. Zimbabwe was a great country in which to study wildlife, and we wouldn't have minded going there anyway, but we looked too much like journalists with our cameras and microphones and would likely get arrested at the first checkpoint. We'd soon be released, of course, but without our equipment and money.

Victoria Falls are twice as high as Niagara; they are claimed to form the largest sheet of falling water in the world. Below them is a narrow canyon. You can swim in a pool right at the top of the cliff, or come at night to see the lunar rainbow, or take a rafting trip through the rapids downstream. We did it all, but we didn't forget about crocodiles. Zambezi River above the Falls was full of them. It would be a nice place to study crocs that lived only in large bodies of water, but I already had such data from Namibia. I was now looking for a place where crocodiles lived only in small bodies of water. That seemed impossible, because it was still the dry season and all crocs were in rivers.

We decided to try national parks farther north and east, and went to Lusaka, the Zambian capital, to ask for permits. By that time we had developed an efficient technique for dealing with officials: Sarit, who could be extremely persuasive, did most of the talking, while Alex and I stood by, looking professional and cutting in if she got too pushy.

This time the initial talk went well, and we were promised an audience with the person in charge. Alex and I returned to Columbus to fix the fan, while Sarit decided to wait in the reception room. She was wearing her standard battle outfit—a tight T-shirt and tiny shorts. The secretary, a middle-aged lady, looked her over a few times and asked:

"What does it feel like, being a white lady like you and dressing like that?"

"Comfortable," Sarit answered. "Our country is very cold, so it's too hot for us here."

The secretary said nothing for a few minutes, then found an audio tape in her desk drawer, and turned on her player. The tape was a recording of a fire-and-brimstone Christian sermon, so emotional that sometimes it sounded outright demonic. After a while, Sarit got tired of listening to anatomically detailed descriptions of tortures used in Hell, and asked:

"How should I dress, do you think?"

"I don't know," said the secretary, "but certainly without showing *these*." She pointed to Sarit's ankles. "Why don't you buy a *chitenge*?" She was referring to a sari-like piece of cloth that most African women wear all the time.

Sarit ran to a street corner, bought a chitenge, and returned to the office. All the female personnel were jubilant; they promised to make sure the official would give us what we wanted and barged straight into his office. Within five minutes we got the permit. I mumbled my usual phrase about studying crocodiles in our spare time, which was, as always, ignored, and we were off.

Zambia wasn't an easy country to travel around. Its highways were so full of potholes that it was better to drive on the shoulder if there was one. Roads leading to national parks were mostly unpaved and extremely rough, or at least seemed so to us. We later discovered another reason for the constant shaking: Columbus was missing a few suspension parts, so even small bumps gave us bone-jarring shocks. It felt like all our dental fillings were about to fall out.

Zambians are beautiful people, very dark-skinned and well built. Like most black Africans, they can be really warm and sincere, but only if you manage to break through the wall of subconscious self-segregation, so that they'd see you as a human, not as an alien creature who should be either feared or milked for money. Sometimes all it takes is picking up a hitchhiker and driving him to his village, but sometimes the wall seems almost unbreakable. Many foreigners live in Africa for years and never manage (or care) to get accepted.

After checking out a few parks, we got to South Luangwa National Park, one of the best in Africa. It was the dry season, so wildlife and sparse tourists gathered along the Luangwa River. The rest of the park, a huge area of dry woodland, seemed almost empty except for a few sable antelopes and duikers. What immediately struck us was the diversity: we saw five species of mongoose on the first night. Despite a very high density of predators, animals seemed unusually relaxed, and we'd often run into entire herds of sleeping elephants or giraffes.

Part of the park was a high plateau, separated from the Luangwa Valley by steep slopes. Creeks flowing down these slopes had numerous waterfalls. I soon realized that crocodiles living on the plateau were isolated by these waterfalls from crocs in the valley. They were smaller and less shy, probably because nobody and nothing ever hunted them in the deep forest ravines they inhabited. They all lived in tiny pools in almost-dry forest streams. This was an ideal "small body of water" study site for me.

The first morning we went out to listen to their "songs," we made a surprising discovery. These crocodiles sounded completely different from the ones in the Caprivi Strip. They gave deep, powerful roars, lower in pitch than those given by lions. Unlike Indian muggers, they never responded to lion roars, probably because lions sounded too weak to them. And they included roars in almost every "song," unlike the big crocs in Namibia that often gave only headslaps.

Was it related to habitat? I didn't know yet, but I was going to find out. I couldn't hope to find enough "large lake" and "small pond" sites to create a chessboard pattern, like I was trying to do with American alligators. Instead, I was hoping to find pairs of sites—one "big lakes," one "small ponds"—in three different parts of Africa. The Caprivi Strip and South Luangwa were my first such pair.

Finding crocodiles in steep ravines in dense forest was difficult. We ended up spending a lot of time in South Luangwa, where we completely ran out of food and had to cook forest fruit. Every time we had a few spare hours, we drove down to the valley to enjoy the amazing array of wildlife along the big river. The most interesting animals to watch were usually the two species with trunks: the elephants and the little, bipedal, incredibly funny elephant shrews that hopped around like tiny kangaroos. Buffaloes, zebras, and hippos lived there in herds many hundreds strong. Probably because of this abundance, valley crocodiles would often leave the water at night to ambush-hunt on game trails. Once we saw a crocodile galloping after a bushbuck (a small antelope). But the reptile covered only about fifty feet before getting tired and giving up the chase.

The downside of being surrounded by so many big animals was having to deal with the myriad tsetse flies. The chances of contracting sleeping sickness from their bites were small, but the bites were very painful. These flies' mouthparts are designed for piercing pachyderm hides. The tsetses are amazingly robust; you can slap them with an open hand or even squeeze inside your fist—and they will keep flying around as if nothing happened. They are particularly interested in large, moving, solid-colored objects and can chase cars at speeds of up to twenty miles per hour. Recently, it was more or less proved that zebras have evolved their stripes to fool the tsetse flies. These flies give birth to huge larvae, one at a time, and even produce milklike liquid to feed them inside the uterus.

On our last day in the park we decided to spend the night in a tiny shed overlooking the river. It was officially called Hippo Hide. In the morning we waited until all hippos returned to the water (their main trail was ten feet from the hide), walked to Columbus, and had just started driving when a pride of lions brought down a buffalo in front of our truck. When we managed to stop, the buffalo was lying under the cabin door.

What followed was awful. Normally, if there's a male in the pride, he is supposed to finish off any large prey. The standard way of killing a buffalo is called "muzzling," where the lion opens his mouth wide, puts it over the victim's muzzle, and waits for it to suffocate. But the male of this pride was very young and seemed inexperienced. Every few minutes he'd get distracted by the taste of blood, take his mouth off the buffalo's muzzle, and start licking the blood off. So the buffalo remained alive. Lionesses and cubs didn't wait; they started tearing the poor animal apart, starting with the softest parts. It took the buffalo half an hour to finally stop moaning and die. By that time it was thoroughly gutted, had its face chewed off, and was lying in a pool of blood. The lions were bloody from head to toe, and even Columbus looked like it had been hunting buffaloes with them.

My friends were deeply shaken. Sarit managed to keep taking photos all the time, but I could see that it was too much for her. We never thought we'd be glad to leave South Luangwa, but we almost were. Later we decided to use a photo of a blood-soaked lioness for the front cover of our book. We wanted it to be about the real Africa, not the one from tourist brochures.

23
Traps on Trails
Crocodylus niloticus

In Africa a thing is true at first light and a lie by noon.
—Ernest Hemingway

THE NEXT COUNTRY ON OUR LIST of potential research locations was Malawi, which stretches along the long, narrow Lake Nyasa (which Malawians call "Lake Malawi"). I wanted to have a study site at one of the African Great Lakes.

Crossing borders was never a routine procedure in Africa. Almost every time, border officials on both sides tried to hit us with all kinds of fees and taxes, most of them locally invented. "Greenhouse gas tax" seemed to be the most popular at the time. I remembered that during my first trip to Africa four years earlier, "AIDS certificate tax" was in fashion. We used our credentials and a bunch of important-looking documents in Russian to avoid paying, or simply drove off before the officials could say anything.

On the Malawi border we were asked to pay a huge fee. Sarit, who, as usual, was doing most of the talking, lost her patience and raised her voice a bit. The immigration clerk got really mad, shouted that by offending him we were offending the Malawi nation, and refused to let us into his country. Alex and I had to wait for half an hour and apologize. Playing on macho men's stereotypes is always easy. "She is a woman," we said. "What did you expect? Of course she doesn't control her tongue. None of them do." That worked.

We got to Lilongwe, the capital, and stopped there for a few days. The town was small and nice, car mechanics were cheap, and the tourism office responsible for issuing permits was very efficient. By that time Jackal Buyback had stopped answering our emails or paying for truck repairs, so we had to buy new springs for Columbus ourselves.

For me, passing through any large city was a long-awaited opportunity to talk to Nastia. Internet access was always ephemeral and unreliable, so I'd pretype an account of our latest adventures and instantly post it on my blog as soon as the connection could be established. Then, if it was still working, I'd try calling Nastia on Skype or chatting with her. Sometimes I could send only an email. Nastia was a bit upset when she had to read the news on my blog instead of hearing it from me personally. She couldn't imagine what it was like to wait for twenty minutes just for the first page of Gmail to download. She was also a bit upset about missing so much fun. No matter how I tried to describe our work realistically, it always sounded easier, less stressful, and more entertaining than it really was.

From Lilongwe we drove down a steep escarpment to Lake Malawi. The lake, over 350 miles long, is a nice place to have a break from all the heat, dust, and bumps of African roads. It fills an ancient rift valley, and, just like two other unusually old lakes—Tanganyika and Baikal—is full of wonderful, absolutely unique creatures. Scuba diving in its clear waters was fascinating. As soon as we got under the surface, we were surrounded by huge shoals of *mbuna*—small cichlid fish belonging to dozens of species, some sky-blue, some bright-yellow, some striped. Vertical rock walls led to the sandy bottom dotted with craterlike cichlid nests. There was even a shipwreck, a rusty tugboat guarded by big, sharklike catfishes.

The only thing missing was crocodiles. I remembered seeing them underwater while snorkeling in Lake Tanganyika and was deeply disappointed to learn that they had been all but hunted out in Nyasa.

We checked croc populations in three national parks further south, but they were all too small. So we decided to go to Mozambique.

We got there through a remote border crossing used mostly by local villagers. A deserted road led to the next town. After a few hours we saw a jeep ahead. It was just a couple hundred feet away when I realized that I had no idea which side of the road I was supposed to drive on. In former British colonies it's usually on the left, in other African countries, on the right. Mozambique is a former Portuguese colony, but it's surrounded by former British colonies on all sides. What should I do? Fortunately, the other driver was so amazed to see another car on that road that he pulled over.

We crossed the mighty Zambezi and kept driving south. Our search didn't go well. Most local crocs had been killed during the recent civil war. One national park had lots of crocodiles, but the park fees were too high and we didn't have a permit for that country.

Columbus slowly kept falling apart. To start the engine we had to pour gasoline into the carburetor. Alex was rapidly becoming an expert mechanic. He even invented a new method of plugging tire punctures with folded duct tape. But he couldn't fix everything. We managed to get the director of Jackal Buyback on the phone, but he just told us to bring the truck back. "We'll make sure you aren't unhappy with our service," he said vaguely.

To break our endless drive, we made a few stops at the coast. We decided that our guidebook would have an extensive chapter on diving, and "sampled the services" of a few dive operators. We were especially impressed with a small town called Tofo, where we got to swim with ten manta rays the size of a small airplane and five whale sharks on the same dive.

The closer we got to the South African border, the more difficult it was to get through frequent police checkpoints. All policemen wanted bribes. Sometimes they'd claim that our papers or license plates weren't in order; sometimes they'd complain about Columbus not having seat belts (we had to take those off because they were attached in such a way that in the event of an accident they'd break our necks). Sometimes they said we'd been speeding (that was usually true, but they never had radars). Sometimes they couldn't find anything wrong but asked for a bribe anyway. Apparently, South African tourists had a tendency to part with their money too easily—even the amounts of fines were given to us in South African rands. We firmly decided never to pay anything. We knew that if we paid once, the policemen would radio to other posts down the road that an easy prey was moving in their direction. Nine times out of ten I'd simply ignore the police and drive straight through. If they had big guns or were standing in a chain across the road, we'd pretend we could speak only Russian, or demand to see the US consul in Maputo (the capital), or waive Obama portraits (he was running for president at that time), or tell them that Sarit had Ebola fever and was about to die on their hands.

Once in South Africa, we learned that according to their rules we couldn't get a second visa in a row without going to our home country first. I was used to visa-related laws and regulations being the pinnacle of human stupidity, but that was way too much. Eventually, we managed to get some kind of emergency visas, but for that we had to pay a lot of money and spend a day in a youth hostel making fake bank statements and house rental agreements on their computer. We had no choice: my ticket out of Africa was from Johannesburg, and we had to return Columbus to Cape Town.

Our next stop was Kruger National Park in the northeastern corner of South Africa. It's one of the oldest, largest, and best-protected national parks in the world. In the rest of Africa, decades of ivory hunting have led to elephants evolving smaller tusks, but in Kruger you can still see old males with ten-foot tusks.

We went to the chief warden to get his permission for our long stay in remote parts of Kruger. It wasn't a good moment. His friend, a ranger, was near death. He had been leading tourists on a guided hike through the bush and walked straight into a lioness with small cubs. He had a gun but was reluctant to use it and missed the opportunity. People working in African parks are often incredibly dedicated to protecting wild animals. They have to risk their lives every day, especially in areas with lots of armed poachers (which is almost everywhere). And they are all very nice people. We got our permission in five minutes.

Kruger is so huge that from north to south it spans five vegetation zones. Most of its wildlife and virtually all tourists are concentrated in the south, where grazing is better and access easier. We had to go to the far north, where crocodiles were waiting out the dry season in small pools left in dried-up riverbeds. Of course, it wasn't boring there either: we still saw lots of rhinos, buffaloes, elephants, and other animals, even rare roan antelopes.

The end of the dry season was still almost a month away. The pools where crocodiles lived were the last watering holes in the parched savanna. So the crocs had a lot of large prey, and grew huge. One male was about fifteen feet long. His roars were so low-pitched that they sounded like distant thunder. One morning I found him holding a torn piece of cloth in his jaws. Did he catch a hapless Zimbabwean refugee during the night? Possibly. A lot of people attempt to cross from Zimbabwe or Mozambique into South Africa through Kruger, and it's thought that at least a few dozen get killed by animals every year, although lions are the main culprit.

Nile crocodiles have undoubtedly hunted people since the age of the first humanlike apes. Although news from small African villages seldom reaches the media or the official tallies, it's estimated that this species kills six hundred to eight hundred people every year. That's five to six times more than all other crocodilians combined. Three out of five Nile crocodile attacks are lethal, compared with one in twenty attacks by American alligators. Nile crocs are not only larger (relatively few gators are more than ten feet long), but they are also more used to hunting big mammals. Interestingly, even an attack by a fifteen-foot crocodile is not always a death sentence. Some people have managed to escape by pressing on the reptile's eyes, or by pushing their hand deep into its mouth and pulling on the valve that closes off its throat underwater. Of course, they have still sustained horrible injuries. Even a single bite by a three-foot-long croc has been known to cause a person to bleed to death.

The most famous man-eating crocodile is Gustave, a large male living at the northern tip of Lake Tanganyika. He mostly stays in Burundi waters but crosses into Tanzania or Congo-Zaire every time yet another hunting party goes after him. For decades he has avoided all kinds of traps. He developed a habit of taunting crocodile hunters before sneaking up on them and killing them. When not hunted, he is surprisingly tame and has even been filmed for TV documentaries. Local people fear him to the point of worshipping him as an evil god. They claim that he has killed more than three hundred people and that he often kills for fun, leaving the bodies uneaten. He is also claimed to be the largest living Nile crocodile, over twenty feet long. Nile crocs living farther from the equator never get that big, perhaps because they virtually stop growing in winter. Gustave is big enough to kill adult hippos—probably a dream of every croc in East Africa, where hippos often bully and sometimes kill crocodiles. I'd really like to hear him roar, but when I saw him on his favorite beach back in 2004 it wasn't during the mating season. Besides, he lives in a huge lake, so his roar probably isn't loud at all.

That was the pattern we were finding. Crocodiles living in small pools and streams had loud roars; crocs living in big rivers and lakes gave only quiet "coughs." We heard plenty of those weak "coughs" in our next study area, Saint Lucia Lagoon on the coast farther south. We got there by crossing Swaziland, a small, neat country with just one national park, called Hlane. The entrance to the park was decorated with a pile of wire snares confiscated from poachers. It was three stories high.

Saint Lucia Lagoon is a river estuary separated from the ocean by a broad sandspit. Crocodiles sometimes swim in the ocean and get killed by great white sharks on occasion. The lagoon, which is part of the recently created iSimangaliso Wetland Park, is one of the southernmost places where Nile crocodiles live. Winters are so cold here that crocs hibernate in underground burrows for three or four months a year. To make fasting easier, they slow down their metabolism, breathing, and heartbeat. In addition, they change the configuration of their heart, making it three chambers instead of four, so that unused oxygen can still be utilized during the next cycle of circulation. Crocodiles also do that to save oxygen while diving. No other living creature has such a flexible and efficient heart.

We didn't see much wildlife in iSimangaliso in the first few days, but then one night, driving through the hills, we spotted a large male leopard. At that time Columbus was being fixed in a nearby town, so we were driving a "safari bus" provided by the park, a small open vehicle with no doors. African leopards are usually shy and difficult to see, but this male walked straight toward us and started circling the car, trying to look in. Alex and I were following him from inside the car with our flashlights, while Sarit was shooting photos. To me, a former cat owner, it was clear that the leopard wasn't in hunting mode—he was just curious to see what's inside. Then he got ready to jump in. That would be bad. Just a month earlier we talked to a safari guide who once had a leopard jump into his open jeep during a night drive with tourists. He pushed the cat out with the butt of his gun, but not before the leopard made a swipe with its paw, and its claws completely tore off the guide's nose and two fingers. So, as much as I'd like to pet a wild leopard, I couldn't let him jump into our car. I grabbed a spotlight and pointed the light beam into his eyes. He blinked and turned away. Then he circled us one more time. I started moving slowly, but he followed us for another minute or two.

After we were done observing crocodiles, we went to another national park, called Hluhluwe-iMfolozi, to rest in cool hills and tally the data. By now we had left the tropics, where the dry season is from April to November, and entered the subtropics, where the pattern is reversed. It was October, so everything was green, all animals had newborn babies, and scarlet *Haemanthus* flowers the size of a volleyball dotted the meadows.

I tabulated my crocodile data and got strange results. Crocodiles in all habitats used a lot of headslaps. But in "small pond" areas they also roared almost every time, and their roars were very loud. In "big water" places they seldom roared, and their roars were barely audible.

That was the opposite of what I had found in alligators. Alligators always bellowed loudly, but they headslapped a lot only in "big water" habitats, and almost never did it at "small pond" sites. One-half of my theory's predictions worked only for alligators, the other half only for crocodiles. I had no idea how to explain this. For now, all I could do was keep collecting data.

But first we had to return the increasingly troublesome Columbus to Jackal Buyback. The worst incident happened during a wild dog chase. We had always wanted to see African wild dogs. They need huge areas to hunt and are highly susceptible to the diseases of domestic dogs, so they survive in very few places and are always rare. In Hluhluwe-iMfolozi we found a pack, but just as we began to follow it and gradually get closer, the carburetor broke down again. We somewhat fixed it, but from that day on we always had to park Columbus on a slope overnight, because in the morning the engine would start only if we rolled downhill. Our windshield wipers, headlights, all gauges, and rear brakes were no longer working, and the exhaust pipe was about to fall off.

Driving around Africa we had met a few other people who had buyback trucks from Jackal Buyback, and we later kept in touch. They all had lots of problems, and reported that contacting the company was becoming increasingly difficult. One American couple, Jim and Nancy, were so mad at Jackal that they considered suing them.

We drove across the Drakensberg, a Jurassic-looking chain of cliffs ten thousand feet high, famous for Bushman rock paintings and some of the world's best hiking trails. A narrow, slippery, very steep dirt road leads up those cliffs to the frigid, extremely arid, Tibet-like plateaus of Lesotho. Then we followed the lush, beautiful eastern coast of South Africa to Cape Town. We arrived there six hours before the scheduled time for returning Columbus. The battery was almost dead, so we had to sleep in the cabin and keep the engine running on our last night with the truck.

Our friends Jim and Nancy had returned their truck a day earlier. They were promised by the Jackal director to be paid more money than they were entitled to by their contract, as a compensation for all the time they had to spend fixing it. The payment was to be made in three business days.

I couldn't wait to get rid of Columbus. Driving in Africa is difficult enough as it is; doing it in a truck that breaks down every few hours and can get you killed any second is exhausting. I was the most experienced driver among us and had a lot of prior experience with marathon drives, so I ended up being behind the wheel almost all the time. In the last few days I was so fed up with it that every time I heard the word "driving" it seemed to me that those hard, dented "dr" consonants were scratching my brain. Also, despite having unusually diverse life experiences, in three months we ran out of stories to tell each other on the road.

We prepared for a hard talk with the Jackal director. We were going to demand reimbursements for all repairs, and a huge discount for all our suffering. In four months we had eight major breakdowns and almost a hundred small ones, not including puncturing tires about fifty times and having to replace them twelve times. Usually such talks were Sarit's job, but this time we all were ready for battle.

We drove to the office of Jackal Buyback. The huge doors were open. Inside it was absolutely empty. No people, no trucks, no computers, no file cabinets, no lightbulbs. They were gone.

We asked the neighbors, but they didn't know anything. One security guard told us that he had seen Jackal personnel packing up and leaving the day before. We were stuck with a near-comatose truck for which we didn't even have a title. Still owing us $12,000, Jackal Buyback had vanished.

24

The Rainy Season

Crocodylus niloticus

If you are among lions, don't look like a zebra.
—Zulu proverb

We went to a police station. The detective on duty listened to our story and said:

"The address of that company is on the border of two police precincts. Our precinct sucks. Nobody will ever help you here. Why don't you go to the other one? They are much better."

We thought he was just trying to get rid of us, but he was probably right. In the other precinct the detectives were very helpful. They soon found that the Jackal director had once filed a complaint about his neighbor's dog misbehaving on his lawn, so he was in their database. Susie, a young policewoman, drove us to the director's home. But the place was locked up, and the neighbors said the family had left in a hurry the day before.

"I'll do what I can to help you," said Susie, "but it doesn't look good. Don't hire a lawyer. They are useless in this country. I own a youth hostel. You are welcome to stay there for a while. I'll give you a discount."

So we did. The next day we found that Jackal had filed for bankruptcy, exactly what my mother had predicted four months earlier, but I have no idea how she could have foreseen the events leading to it: the worldwide economic crisis and the sudden fall of the South African rand.

We called Hyena Overland, a large tour company. Jackal was their subsidiary. But they said they had nothing to do with its debts. All they could tell us was that Jackal could give us the title for Columbus once the bankruptcy proceedings were over. That wouldn't happen for a month. It was very obvious that they expected us, and all other Jackal customers stuck with leased trucks, to have to leave the country by that time.

They were wrong. I had a return flight in three weeks, but Sarit and Alex weren't planning to leave Africa any time soon and could easily stay in Cape Town for a few months.

For now there was nothing we could do. We decided to go to Madagascar to look for its cave-dwelling crocodiles. The peculiarities of African ticket pricing meant that we had to fly to Durban, rent a small car, drive to Johannesburg, and fly to Madagascar from there. Susie allowed us to leave Columbus in her yard.

We drove from the coast up to the central plateau of South Africa, a land of beautiful grasslands. The entire continent was celebrating Obama's election. Everybody in Africa considered him a Kenyan. In Kenya, the entire country stopped working for a week. Kenyan TV was showing a musical called *The Life of Barack*, with Kikuyu actors playing Hillary Clinton and John McCain.

In Pilanesberg National Park north of Johannesburg we stopped at a campsite where a few white Zimbabwean refugees were staying. They all had had to abandon their businesses and move to South Africa, where they jointly opened a small safari company. They invited us to their campfire.

Whites who have lived in Africa all their lives are usually very interesting people to talk with. They can tell you things that foreigners don't know and that black Africans consider either too mundane or too sensitive to mention. This campfire conversation was no exception. At some point we were discussing local politics, and someone asked why Mbeki, the president of South Africa at the time, appeared to support Mugabe.

"He doesn't really support him," explained the oldest Zimbabwean. "He's just afraid to criticize him. Everybody in Africa is afraid to criticize Mugabe."

"Why?"

"Because everybody knows that he has excellent witch doctors. Once the president of Zambia said something bad about Mugabe—and in a few days he fell ill and died."

"But the president of Kenya criticized Mugabe just yesterday."

"Oh, but the president of Kenya is in a league of his own. His witch doctors just got a Kenyan elected the president of the United States. They are above all competition."

The next morning we left Africa and flew to Madagascar. We landed in the island's capital, Antananarivo, and tried to book a local flight with the national airline, affectionately known in travelers' circles as Air Mad. And we discovered that all our credit cards had been frozen *again*. If you travel in Africa, expect your bank to block your card every time you use it (even if you have called the bank in advance and told them about your travel plans). It doesn't prevent fraud, which also follows almost every time.

Alex had to call his bank. Making phone calls from Africa is almost hopeless when you have to talk to a voice-recognizing answering machine, because the sound quality is so bad. But Alex eventually managed to reach a human operator. It didn't make things much easier.

"Hello, Visa Express? I had my card frozen. Also, someone just bought a ticket from Pakistan to New Zealand using my card.... No, I did buy a ticket to Antananarivo, that's OK ... and to Durban ... and to Maroantsetra.... No, not to Kolkata, please cancel that transaction.... Yes, please cancel that card and mail me a new one— Wait a second, I can't hear you, there's an airport announcement— Yes, I'm in an airport.... Where to mail it? To Madagascar ... It's a country.... No, it's not a cartoon.... No, I'm not joking.... OK, I'll spell it: Mary-alfa-delta-alfa— No, I don't have an address here, mail it to the American Embassy, I'll pick it up from there— No, it's not a company, it's a place where the ambassador works.... OK, forget it, mail the card to

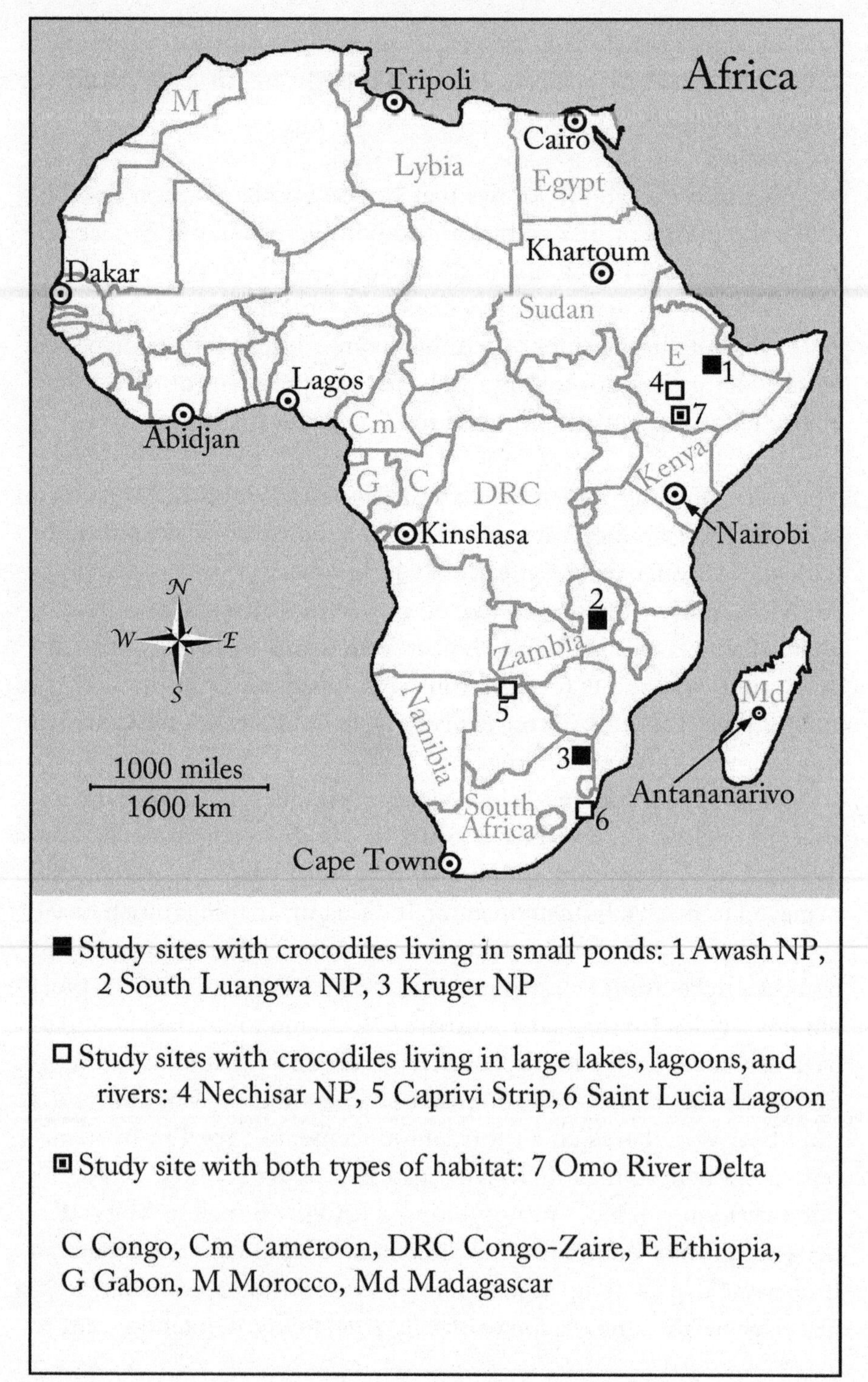
Africa
Tripoli
M
Cairo
Lybia
Egypt
Khartoum
Dakar
Sudan
E
1
4
7
Lagos
Abidjan
Cm
G
C
DRC
Kenya
Nairobi
Kinshasa
N
W
E
S
2
Zambia
Md
5
Namibia
3
1000 miles
1600 km
Antananarivo
South Africa
6
Cape Town
■ Study sites with crocodiles living in small ponds: 1 Awash NP, 2 South Luangwa NP, 3 Kruger NP
□ Study sites with crocodiles living in large lakes, lagoons, and rivers: 4 Nechisar NP, 5 Caprivi Strip, 6 Saint Lucia Lagoon
▣ Study site with both types of habitat: 7 Omo River Delta
C Congo, Cm Cameroon, DRC Congo-Zaire, E Ethiopia, G Gabon, M Morocco, Md Madagascar

South Africa, I'll be there in a month. . . . Yes, it's another country. . . . No, it's not part of Africa. . . . I mean, it's a part of Africa, but it's also a country. . . . I told you already, I'm not joking— Send it to Aardvark Backpackers— It's a hotel in Johannesburg— Aardvark: alfa-alfa-Robert— Wait a second, there's another announcement. . . . Johannesburg is a city. . . . Oh, forget it, don't mail it now, I'll call you later!"

Madagascar is one of my favorite countries. It's extremely poor, the infrastructure is among the worst in Africa (sometimes it takes a week to drive a hundred miles), and the environment is in really bad shape. But once you get to a protected forest, you start seeing some of the world's most amazing creatures at an unparalleled rate. Chameleons ranging in size from one inch to two feet, katydids the size of your fist, giant moths with "tails" almost a foot long, hedgehog-like tenrecs "talking" to each other with their quills, terrestrial leeches glowing green, cougar-like fossas with tails longer than themselves. . . . You never know what the next hollow tree or pile of leaves is hiding.

Biologists call Madagascar "the seventh continent" because it's so unique and diverse. It's separated from Africa by a strait just three hundred miles wide, but the current in that channel is very fast and makes the crossing difficult. Only a few nonflying animals have been able to get to Madagascar from Africa. Fifty-something species of lemurs have all evolved from one lucky ancestor; the same is true for the island's snakes, chameleons, and carnivorous mammals. This crossing was so difficult for humans that the first people to populate Madagascar didn't come from Africa. They arrived more than a thousand years ago from what today is Indonesia, crossing the Indian Ocean.

Crocodiles, however, crossed the strait easily. The first ones were weird horned crocodiles. When humans arrived and killed all large animals on the island—the panda-sized lemurs, the elephant bird, the mountain hippopotamus, and numerous others—those horned crocodiles died out. Then the more adaptable Nile crocodiles arrived. In the last few centuries they have colonized almost the entire western side of Madagascar and are now spreading in the east. They don't grow longer than twelve feet here, but still manage to kill people sometimes.

December is the best time to visit Madagascar. The weather is reasonably comfortable everywhere. Lemurs have babies, it doesn't rain all the time in the eastern rain forests, and the bizarre thorn forests of the southwestern desert turn green literally before your eyes after a thunderstorm.

Sarit and Alex were dying to see Madagascar, and I was happy to return there. But we had very little time. We found only one crocodile, and it didn't "sing." Later I got some information from a researcher working at a crocodile farm in Morocco where all animals were from Madagascar. Apparently, the "songs" of those crocs aren't different from the ones in Africa.

I now had records from two pairs of study sites: one in Southern-Central Africa, the other in the far south of the Nile crocodile's range. To show that the patterns I found weren't climate-related, I needed to add one more pair, preferably from the north of the continent. Crocodiles are almost extinct in Egypt and the Sahara and completely gone from Palestine. Working in Sudan or Mali would be extremely difficult. So we decided to meet in Ethiopia in late February.

For now, I had to leave Sarit and Alex in Madagascar and get back to the US. I had some things to take care of at the university—and besides, I couldn't wait to meet Nastia again.

I flew back to Johannesburg. I had a few hours before my flight to New York, so I went to a small nature reserve near Pretoria. I couldn't recognize the land. While we were on Madagascar the first rains came to the savannas, and the plateau looked like a well-watered lawn. I saw more snakes in one day than we had found in five months. Dozens of them were crossing the road: spitting cobras, beautiful green boomslangs, and lots of puff adders. Every time I see those fat, slow vipers, I remember the book *I Dreamed of Africa* by Kuki Gallmann, and feel happy that it has never been translated to Russian, so my mother will never be able to read it. It's an account of the author's life in Kenya, culminating with the tragic loss of her only son to a puff adder bite.

Back in New York, I picked up my car and drove straight to Tennessee. North America in December looked a bit like Africa during the dry season: gray hazy skies, naked trees, dead grass. . . .

Nastia later confessed that she was seriously scared when she saw me. I'd lost forty pounds in five months in Africa, mostly because there often wasn't enough time to eat. But I got better after a few weeks. I discovered that Nastia was a cooking genius, and though she didn't have kudu meat to barbecue, she still managed to create better dishes than anything I'd ever tried in the ultra-luxurious lodges of Africa. But even if she hadn't been such a great cook, anything prepared by her would probably seem delicious to me.

Sarit and Alex returned to Cape Town two weeks later and settled in Susie's hostel. Alex made a new website for Susie's police precinct; Sarit worked at the hostel reception. They got the title for Columbus and considered driving it to Ethiopia, but soon realized that fixing it would cost more than buying a new truck. Many other ex-customers of Jackal Buyback were in a worse situation: they had returned their trucks before the company's bankruptcy, and never got their money back, so they were left completely empty-handed.

Alex did a lot of online detective work and found that Hyena Overland and Jackal Buyback weren't as unrelated as the Hyena people claimed. The wife of the Jackal director was a close friend and longtime business partner of the CEO of Hyena. Alex posted some information about it on a few travel forums and created a website called "Hyena Overland Fraud," where he told the whole story and warned people to avoid dealing with that company.

It worked instantly. Less than twenty-four hours after it went online, before the website even had time to show up in Google search results, the CEO of Hyena called Alex and asked for a meeting. Alex refused to go to their office and told them to come to Susie's hostel. He wanted to have a uniformed police officer at his side during the talks.

I wish I had been there to see it! After a lengthy argument, the Hyena people agreed to return almost all the money Jackal owed us. They also agreed to reimburse all of Jackal's other customers for their losses. All they wanted in return was the removal of the new website.

Unfortunately, a side effect of our victory was that Hyena Overland decided never to provide buyback services again. Nobody else does it either. So the only way to get a truck for travel around Southern Africa is to rent it or buy it, which is very expensive. Of course, if you stay within South Africa and Namibia you almost never need a 4 x 4, and small passenger cars are really cheap to rent there.

25

Dangerous Crossings

Crocodylus acutus

Man noh done cross riva, noh fi cuss alligator long mout long.
(A man who hasn't crossed the river yet shouldn't call the crocodile "Big Mouth.")
—Jamaican proverb

Living with Nastia in her small apartment in Knoxville felt strange at first. I still had dreams about Africa, mostly involving endless drives, walks, and crocodile watches. When trains would pass within a mile of our house in the quiet predawn hours, I'd often wake up, expecting to see elephants grazing around the tent.

As soon as I put on a few pounds and looked normal again, I drove to Florida to meet with Steve, my advisor. Steve seemed optimistic about my research. I wasn't. I had a huge pile of data, but the results were contradictory and made no sense.

I no longer had my nice apartment in Little Havana; all my stuff was in my friend's shed. I would still go to Little Havana for things like dental services and car repairs, because speaking Spanish could get me a lot of discounts there. I always wondered why Americans living in Miami never bothered to learn a few words of this beautiful language. Anyway, during my visits to Florida I now had to live in my office, an ugly concrete room with steel furniture. The light there came from a corridor through a glass wall and couldn't be turned off. The office looked remarkably like a KGB interrogation room. I saw a few of those when I lived in the Soviet Union in the 1980s, mostly after being arrested for what I called "geographical crimes"—entering various restricted areas without a permit in search of rare wildlife. Sometimes I'd get myself arrested in such areas even if I did have a permit, because that was the fastest way to get back from remote places. The arrests never had any serious consequences. By that time the much-feared KGB was so dysfunctional that it didn't notice the suspicious pattern.

I had less than a month left before going back to Africa, and during that time I had to come up with some way to figure out why my results looked so strange. I also had to get more data on the American crocodile, because winter was its breeding time. I decided that thinking about scientific problems would be more effective in a more relaxed place and took Nastia to Jamaica. That island wasn't too far away, it was small, it had warm beaches—and crocodiles.

Comparing a country to a set of Russian dolls is a common cliché in travelogues, and it's usually a poor description—not to mention that "Russian doll" is a misnomer (they were invented in Japan). But Jamaica fits that image perfectly. The island consists of four concentric rings. On the coast are crime-infected cities, busy highways, beach resorts, and terrifying multitudes of tourists unloaded by cruise ships. Just a mile inland is the world of sleepy towns, narrow roads, orchards, and ranches. When people there see a white person, they are often so surprised that they can't help exclaiming, "Jake!"—the local slang word for a Caucasian. Once you begin climbing into the mountains, you get to a very green and charming country. It's sparsely populated by the descendants of runaway slaves who once managed to create a guerrilla republic there. They are proud, beautiful, and very hospitable people. Few roads penetrate this land. If you manage to get even deeper, into the cool, misty heart of the island, you find yourself in uninhabited cloud forests, full of wonderful creatures like giant predatory cuckoos and snails with shells the size of a small plate. This innermost ring surrounds the barren summit of Blue Mountain Peak, the highest peak.

We landed on the northern coast, but almost all crocodiles lived on the southern side. We took the most scenic road across the mountains, and after driving along countless switchbacks in dense fog got to a place where the road had been totally swept away by a huge landslide. It was almost midnight. We didn't want to drive all the way back and considered sleeping in the car. Suddenly, we heard voices and were instantly surrounded by a crowd of locals trying to look in.

They were really excited to see us. They said we were the first tourists to get there since the road was destroyed three years earlier. All local businesses were in various stages of bankruptcy. They fed us, showed us a school yard where we could park our car near a warden's cabin, and guided us across the landslide to the other side, where a small inn was still open.

The next morning we looked out of the window and saw forested mountaintops rising from a shining white sea of clouds. We were on the highest road pass in Jamaica.

Like most people in the remote interior of the island, our hosts were Rastafarians. They all had impressive dreads down to the waist. Once a day the entire family would gather around the table to smoke "herb" and drink locally grown coffee. In every room, photos of reggae stars were hanging side by side with iconic portraits of Ethiopian royalty. Adherents of this religion believe that the last Ethiopian emperor, Haile Selassie I, was a reincarnation of Jesus. It makes no sense whatsoever to anyone familiar with Ethiopian history, but somehow the Rastafarians managed to develop a faith based on an extremely shaky (and originally racist) New Age foundation into a spiritual system that is remarkably healthy, at least compared with other Abrahamic religions. Rastafarians are much more sincere and open-minded than the buttoned-up Anglicans living on the coast. My Jamaican girlfriend Kami, who went to a Sunday school when she was a kid, has never been given a chance to read Song of Songs—it was considered inappropriate reading by local Christians. Rastafarians are still a small minority in the otherwise deeply Christian country, but they seem to have a considerable influence on the nation's psyche. For example, *Babylon*, the Rasta term for the corrupt Western society (as opposed to *Zion*, the good old society of Africa), is routinely used in everyday Jamaican speech, where it means "police." The local term for female sex tourism, for which Jamaica is a major destination, is "rent-a-Rasta." I always enjoy listening to English-based languages of black West Indians: they are so expressive!

We had to make a big detour to cross the island farther west and get to the southern side. There's a lot to see in this small country, but the best place is a small bird sanctuary hidden in the hills. Among its many exotic inhabitants are doctorbirds, gorgeous hummingbirds with tails three times as long as the rest of their bodies. If you fill a bottle with sugar solution and put an index finger in front of the bottle opening, a pilotbird would hover in front of your face for a few seconds, look you in the eyes, then land on your finger and drink from the bottle.

Once across the mountains, we were in the so-called Cockpit Country, a strange landscape of limestone hills that looked like chicken eggs in a box. This kind of terrain is very difficult to penetrate and even more difficult to build roads through. A large hilltop village called Accompong is the former capital of the republic of runaway slaves. We happened to get there during a colorful annual festival, of which Nastia's arrival was apparently the main event. We were the only whites there, and on top of that she was very blond and light-skinned. Everybody treated her like a plush toy that was sweet, tender, and able to talk. Girls would stay in line to hug her. Even grim-looking old hard-core Rastas gave her warm smiles. We could only thank the popular tourist guidebooks. These books describe Jamaica as a dangerous, racist, and overall unfriendly place. If not for them, all those wonderful mountains would be swarming with tourists. As it is, the crowds mostly stay in walled-off beach hotels.

We descended to Black River Great Morass, the country's largest swamp. It contains the main population of American crocodiles (locally called "alligators") on the island, but even that population is very small. The crocs are very shy, probably due to poaching. I soon figured out that they mostly lived in bowl-like circular depressions reminiscent of "gator holes" in the US and most likely dug by the crocs themselves. In April, at the end of the dry season, these holes would probably turn into isolated ponds, providing water for local wildlife. But in mid-winter the swamp was over a foot deep. Most crocodiles I managed to find were juveniles. Only one male was about eight feet long, and I saw him head-slap a couple times.

By that time Nastia had begun to complain about the way our trip was run. She was tired after having to pass the last of her qualifying exams just a few hours before leaving home. And she wasn't used to getting up before dawn and spending the whole day on the move. Besides, I realized that Great Morass was not a very safe place to be. It was too close to Kingston, "the murder capital of the Americas," and was full of shady characters, the only people not afraid to enter the swamp. Some were growing marijuana in the swamps; others were practicing the local form of witchcraft, called *guzumba* or simply *science*.

So I gave up on Jamaican crocs, and we spent a day on a beach before flying home.

A day after returning to Knoxville I had to leave Nastia and drive back to Florida to catch the first week of the American crocodile's mating season there before flying to Africa. The problem was that I had already observed crocodiles in all easily accessible places.

There was one juvenile croc in a lake on the university campus, and I had been waiting for it to begin "singing" for three years. But just when it was getting big enough some teenager killed it. I don't know why he did it. The American crocodile was still on the endangered species list at that time, so this idiot ended up in prison. The lake has been empty ever since.

The only accessible area with lots of crocs that I hadn't been to yet was Cape Sable, a long sandy beach that forms the southwestern tip of peninsular Florida. It was just seven miles from the end of Flamingo Road, but the trail had been closed since the last hurricane. So I paddled to the cape along the shore in my kayak.

The place was surprisingly busy. It was easy to get to if you had a boat, so campgrounds were packed with fishermen. Park rangers had told me that it was an important crocodile nesting site, but I couldn't find any crocs or their tracks until I explored Ingraham Lake, a large saltwater lagoon on the inland side of the beach. There I managed to record "songs" of two large males.

It was time for me to get back, but I couldn't. The easterly wind was so strong that my inflatable kayak refused to move into it. I'd walk out, but I didn't have a backpack and couldn't carry all my stuff in my hands. So I tried to paddle along an old canoe trail through the mangroves. It was only ten miles long but was completely overgrown. If I tried to get ashore and carry the kayak around the most difficult stretches, I immediately found myself waist-deep in sticky mud between mangrove roots. Soon the kayak bottom got punctured, so I had to stop every half an hour to pump air into it. Since I couldn't find any firm land, I had to press the foot pump with my hands. I was out of food, fresh water, and mosquito repellent. The worst thing was that every minute I spent there was cutting into my last weekend with Nastia before leaving for Africa. And it took me eighteen hours to return to my car.

I stopped at my office to wash myself part by part in a lab sink (there was no shower) and drove straight to Tennessee. It was a boring sixteen-hour drive along flat freeways. As almost everywhere in the US, the speed limits were totally inadequate for such excellent roads, but I managed to get only one ticket that time.

Driving through the dry pinelands of Georgia, I tried to come up with a solution to one major problem with my results.

I already knew that the "songs" of Nile crocodiles and American alligators differed between animals living in large and small bodies of water. But I also knew from the results of my studies in Brazil and Bolivia that caimans didn't change their "songs" when the size of their lakes changed. Why such a contradiction? I thought that those differences were the result of slow evolutionary change, rather than of individual responses by the animals. But there were two alternative explanations. First, I could simply be wrong to combine data from different species. It could be that crocs and gators did have those differences between habitats, while caimans didn't. Second, it was possible that the animals didn't change their "songs" immediately but took time to do so. I gave caimans only one week to adjust to changes in habitat. What if they needed months, or even a full year?

I had to design a study to figure out which of the three explanations was the right one. And by the time I crossed the Tennessee state line, I had an idea.

So far, in all my studies of Nile crocodiles and American alligators I specifically looked for geographical areas where all animals lived either only in small ponds or only in big lakes and rivers. What if I picked a location where a few small pools were very close to a large lake? In both species, juveniles disperse from the place of their birth as soon as they no longer need their mother's protection and usually move at least a few miles away. But once they get big, they tend to stay in the same place for many years, particularly male crocs, which are very territorial. So in an area with such a mix of habitats, all animals would be genetically close, but most of them would've spent much of their adult lives in either small ponds or big lakes. If their "songs" still didn't differ, it would support my "evolution only" explanation. If their "songs" did differ between habitats, it would mean that one or both of the alternative explanations were correct. In that case, another study of caimans would give me the final answer.

When I got to Knoxville I looked all over Ethiopia on Google Earth, trying to find a place where crocodiles could live in a big lake and in small ponds nearby. There were lots of big lakes, but most of them were surrounded by steep cliffs, mountains, and dry deserts. I found only one place that looked just right. It was extremely remote. What if I cross all Ethiopia to get there, only to find that it has no crocodiles or the lakes are inaccessible or something else is wrong?

But I had no choice. In other parts of Africa the crocs' mating season was already over. It had to be Ethiopia.

26
The Place We Are From

Crocodylus niloticus

Don't blame God for creating a crocodile—thank Him for not giving it wings.
—Oromo proverb

I WAS GOING TO AFRICA FOR LESS THAN THREE MONTHS, but leaving Nastia behind felt much worse this time. We'd grown very close in the few weeks we'd spent together. My only consolation was in knowing that Ethiopia would be too difficult for her. Nastia didn't have enough experience in extreme travel yet.

Flying abroad from Knoxville, with its small airport, was always too expensive. This time it was cheaper to rent a car, drive to Washington, DC, and fly to Addis Ababa from there. During the long flight I tried to distract myself from sad thoughts by working on the Africa guidebook. By that time its Russian title had evolved into *Wild Africa: The Naturalist's Handbook*. I knew that its Ethiopia chapter was going to be interesting.

Ethiopia is the most unusual country in Africa. Its people are black-skinned, but, except for a few border tribes, their facial features look Middle Eastern rather than African. The main languages belong to the Afro-Asiatic family, which also includes Hebrew, Aramaic, Arabic, and the languages of Ancient Egypt and Babylon. The country's heartland, the highlands, has been Christian since the fourth century AD; it's the only ancient Christian enclave in Africa other than Egypt's Coptic community.

But what makes Ethiopia unique is that it's the only country in Africa that has never been colonized. In 1936, despite suffering two humiliating defeats by the Ethiopians, Fascist Italy managed to occupy it for a few years, but the only lasting effect was the introduction of pasta, now very popular in cities. The northern part had experienced a much longer Italian occupation, and the resulting cultural differences were so deep that it later broke off and is now an independent country, Eritrea.

Before the Atlantic slave trade and the colonial era, Africa had survived a much worse invasion. The armies of Islam conquered almost a half of the continent, and Arab slave traders exploited much of the rest. Ethiopia was the first African country to get attacked, but the Christians of its central highlands managed to defend themselves through almost a thousand years of nearly constant warfare.

You'd think that this lack of a colonial past wouldn't matter much. There are plenty of remote places in Africa where neither Muslim nor Western invaders have ever set foot. African cultures often seem remarkably well preserved. But there's a difference between Ethiopia and the rest of Africa. It is in everything: in the unusual menu of roadside restaurants, in the language of interethnic communication (which is Amharic rather than Arabic, English, French, or Portuguese), but most of all in the eyes of local people. They aren't afraid of you. They can be friendly or not, they can beg you for money, offer you food and shelter, ask you double the price for a hotel room because you are a foreigner, or even throw a rock at your windshield, but they don't consider you fundamentally different. That wall between blacks and whites, almost impossible to break through in other countries, doesn't exist in Ethiopia.

And that's what makes Ethiopia so difficult for many visitors. We can complain all we want about the horrors of globalization and the loss of ethnic diversity, but in reality a truly original and alien culture is too much for most of us. We want to see it colorful but tamed and adapted.

The Ethiopian backcountry is dirty, cruel, sometimes violent, and always fascinating. There are no other places like that left. Tibet, Afghanistan, New Guinea, Congo, and the Amazon have all been "civilized." Ethiopia will soon follow, but for now it's your last chance to try traveling the way it used to be done when our entire planet was wild, untamed, and dangerous; when explorers went to distant lands expecting to face something completely alien and unpredictable. Would you be able to leave the harbor like they did, knowing that everything will be for real? The rocks, the arrows, the diseases, the necessary sacrifice of your health and probably your life. And the discoveries will be for real, too.

Addis Ababa has a very pleasant climate; it's never too hot or too cold. The city has only a few parks, but it's full of interesting birds, from sunbirds to ibises, probably because local people almost never keep cats. I stayed in the oldest hotel in all of Ethiopia, with authentic nineteenth-century plumbing, and crisscrossed the city trying to obtain all the necessary permits, find a car to rent, and see all the museums and cathedrals.

The main problem was the car. Ethiopia has a 100-percent importation tax, and you are required to rent a 4 x 4 with a driver if you want to leave the capital. So the rental costs are high even by African standards.

After running around all day, I spent my evenings in Internet cafés trying to read my email. The Communist government of Ethiopia had fallen seventeen years earlier, but the telecom services were still largely a state monopoly, so it took up to an hour to download one page of plain text. I used this last chance to communicate with Nastia because I knew things would get only worse outside Addis.

The city was full of art—beautiful music, old architecture, ancient paintings. In church frescoes, two recurrent motives were particularly interesting. One common image was of King Solomon and the Queen of Sheba, caressing each other in their marital bed. The Ethiopian Imperial dynasty claimed descent from that royal couple, even though there was no real bloodline, because the throne had been usurped by commoners on numerous occasions. The second popular theme was the last emperor, Haile Selassie I, talking in the League of

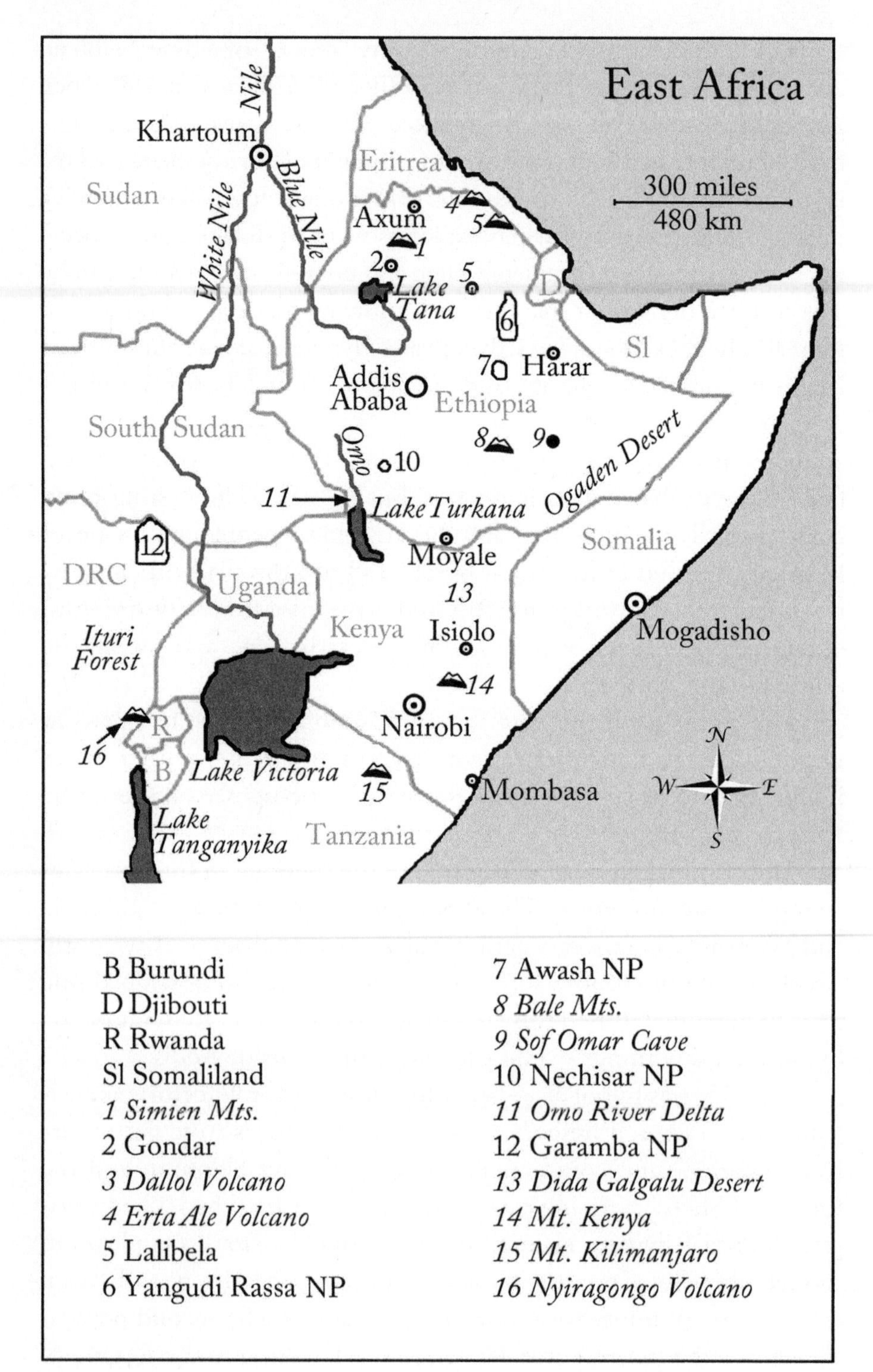
East Africa
300 miles
480 km
Nile
Khartoum
Sudan
White Nile
Blue Nile
Eritrea
Axum
Lake Tana
D
Sl
Addis Ababa
Harar
Ethiopia
South Sudan
Omo
Ogaden Desert
Lake Turkana
Somalia
Moyale
DRC
Uganda
Kenya
Isiolo
Mogadisho
Ituri Forest
Nairobi
R
B
Lake Victoria
Mombasa
Lake Tanganyika
Tanzania
N
W
E
S
1
2
4
5
5
6
7
8
9
10
11
12
13
14
15
16
B Burundi
D Djibouti
R Rwanda
Sl Somaliland
1 Simien Mts.
2 Gondar
3 Dallol Volcano
4 Erta Ale Volcano
5 Lalibela
6 Yangudi Rassa NP
7 Awash NP
8 Bale Mts.
9 Sof Omar Cave
10 Nechisar NP
11 Omo River Delta
12 Garamba NP
13 Dida Galgalu Desert
14 Mt. Kenya
15 Mt. Kilimanjaro
16 Nyiragongo Volcano

Nations. He delivered a passionate speech there, asking it to protect his country from the Italian invasion. The League responded by imposing an arms embargo, knowing perfectly well that it would affect only Ethiopia, since Italy had its own weapons industry. But that speech had an unforeseen consequence. The emperor made such an impression worldwide that in Jamaica a new religion of Rastafarianism was born; one of its central beliefs was that Haile Selassie I was Jesus reincarnate.

As soon as I managed to rent a car for a price that wouldn't require me to starve, I asked the driver to take me to the far southwest of the country, where the poorly demarcated and still-contested borders of Kenya and South Sudan meet in the salt desert near Lake Turkana, the northernmost of the African Great Lakes.

Millions of years ago the African continent began to break up. Cracks ran from Syria in the north to Mozambique in the south. The deepest of these rifts are now the Red Sea and the Gulf of Aden, which separate Africa from Arabia. Other cracks contain large lakes, from the Sea of Galilee and the Dead Sea to Lakes Tanganyika and Nyasa. The rifts are still growing and are lined with volcanoes and hot springs. They are known collectively as the Great Rift Valley, although in reality there are many valleys, sometimes parallel to each other or forming three-way junctions. Some of Africa's most scenic places—Ngorongoro Crater, Ol Doinyo Lengai, the Virunga Volcanoes, Murchison Falls, and the Rwenzori Mountains—are located along the Great Rift Valley.

We left Addis Ababa and descended from the highlands into the Rift Valley. My driver, Tesfaye, was relaxed and in good spirits. He had no idea that a crazy time in his life had just begun. Like many middle-class Africans, he was a recent convert to Protestantism. American churches maintain a lot of missions in Africa; you need to be a person of a certain means to be allowed to join a Protestant community, but it can help you a lot since you get to know all the right people. Being a neophyte, Tesfaye was full of religious fervor and tried to convince me that the Rift Valley couldn't be more than five thousand years old.

We stopped for dinner. Ethiopian food is totally unique. The staple is *injera*, a huge crepe that is supposed to be made of *teff*, a local kind of cereal. You can get a real teff injera only in or near the highlands of Ethiopia; in more remote parts of the country and in all Ethiopian restaurants abroad injeras are made of wheat or barley. Injera is food, dinner table, and eating utensil at the same time. Other dishes (stews, salads, spicy pastes, cubes of raw meat) are served in piles on its surface, and you are supposed to tear small pieces off its edges and use them to pick everything up with your hand. Ethiopian cuisine takes some getting used to, but it's certainly tastier than the stuff you get in roadside eateries in other parts of Africa.

The road mostly followed the Rift Valley, occasionally climbing into the surrounding mountains. One town had a small Rastafarian community. After decades of petitioning the Ethiopian government, the Rastafarians were finally permitted to "return" to their promised land, but so far very few have jumped at the opportunity.

We spent a night in a hotel that had both bedbugs and fleas. Fortunately, I had plenty of mosquito repellent. I shared it with Tesfaye and gave him some pills for his stomach. The raw meat didn't go well with him. I was very proud of myself: being able to eat food that makes even local people sick was the ultimate dream of any tropical biologist. Tesfaye was so grateful that from that day on he allowed me to drive sometimes, even though he wasn't supposed to.

Ethiopia is probably the worst place to drive, at least in Africa. Almost all drivers have only recently learned their trade, and almost all locals aren't yet aware of the danger. Paved roads are rare and particularly scary, because everybody is so happy to see pavement that they drive at maximum speed. There are, of course, the usual African hazards: foot-deep potholes, dust clouds, cattle, children, and flying gravel. A popular belief is that if you run across the road in front of a fast-moving car at the last possible moment, your life will get longer. Another charming popular custom is to push a senile goat or sheep in front of a car. If the animal gets killed, the entire village instantly gathers with pitchforks to make sure the driver pays the liberally estimated "market price" to compensate the owner.

We turned off the highway and drove west, toward Omo Valley. This small area of Ethiopia is often compared to an ethnographic museum: It is inhabited by forty tribes speaking languages belonging to five different families. The best known are the Mursi, famous for their custom of stretching their women's lips into huge loops, sometimes big enough to put a dinner plate–sized disk inside. The larger the disc, the more beautiful the woman, and the more cattle it takes to buy her as a wife. Other tribes also modify their appearance. Ethiopians in general are very beautiful people, but this country has the largest variety of customs involving mutilation. Even the relatively "civilized" highland Christians practice carving or deeply tattooing crosses on their children's foreheads, as well as male and female circumcision, the latter often of the most extreme kind.

The road turned from gravel to dirt to sand. After a long day of driving we reached the bleak shores of Lake Turkana, a.k.a. Lake Rudolf, a.k.a. the Jade Sea, the world's largest alkaline lake.

This place is very hot, dry, and windy. There are volcanic islands and rocky cliffs in the Kenyan part of the lake, but the small Ethiopian part is shallow, boring, and surrounded by low dunes and salt marshes. I tried snorkeling in the lake, but had to walk for half a mile to get knee-deep. The water of Turkana can be drunk only under extreme circumstances. Many of its fish species prefer to spawn in Omo, the only large river flowing into it.

Except for an occasional gerbil, lark, lizard, or carpet viper, there were no wild animals on the shores. The mudflats had a lot of shorebirds, and sometimes I saw a flock of flamingos flying far offshore, or a huge turtle surfacing to breathe. Judging from the number of crocodiles, there were plenty of fish in the lake. The local crocodile population was once Africa's largest, and even today there are said to be about ten thousand. There were lots of them in the shallows, usually in groups of up to a dozen.

It was hard to believe that this inhospitable land was part of the area known as the Cradle of Humankind. Dry, barren stretches of the Rift Valley in Kenya and Ethiopia have yielded most of the oldest hominid remains. Lake Turkana fossils are two to four million years old. At that time, the local climate was more humid, and at least three different species of humans lived here. I once found a primitive-looking stone scraper, but it failed to convince Tesfaye that human history was more than a few thousand years long.

Omo River was half-dry, but in its delta there were numerous shallow ponds left in the mudflats, and many of them contained groups of crocodiles. This was a perfect setup for me: Lake Turkana and those tiny ponds were all within an hour's swim for a crocodile, but during the dry season they were isolated by mudflats. The mud was deep, soft, and sticky, so the crocs never tried to cross the mudflats; I didn't see a single crocodile trail connecting the ponds.

Alistair Graham, a Kenyan biologist, studied Lake Turkana crocodiles in the 1960s, and wrote a nice book, *Eyelids of Morning,* about them. He was one of very few researchers to pay attention to crocs' mating displays and left a detailed account of headslaps but never mentioned roars. I knew what to listen for, so I was sometimes able to discern a weak "cough" in the split second preceding the slap. Graham also described crocodiles roaring in aggressive encounters. Indeed, I saw one fight between two large males in which both crocodiles roared like powerful motorcycles. So they were perfectly capable of roaring, they just didn't use it in their regular "songs."

But what about the crocs in the small ponds? To observe them, I had to crawl across the mudflats for up to five hundred yards. I had to leave all clothes in the car, because mud stuck to them even more than to skin. Fortunately, the local Daasanach people who herded their cattle in the surrounding desert didn't mind. They walked around naked except for woven miniskirts. Some were completely naked; I later learned that these were low-caste men who had no cattle and survived by hunting crocodiles and fishing, and women who for some reason hadn't had their clitoris cut off in childhood and so weren't allowed to be called women, marry, or wear clothes. All men had AK-47s and short spears—they'd kill any animal they could for food and any member of a different tribe for trespassing.

The Daasanach weren't particularly hospitable, probably because all tribes in the Omo Valley were constantly at war with each other, so there was no tradition of hospitality to strangers. We did get invited to their movable dome-shaped huts a few times, but communication was difficult because they spoke very little Amharic, and Tesfaye didn't know any local languages.

I soon taught Tesfaye to watch the crocodiles' behavior, listen for the "coughs," and record the data. Now we could work together. He refused to crawl across the mud, so I let him have the more easily accessible crocs. The midday heat was often unbearable, but the crocs would only "sing" in the morning, so we could spend the hot hours soaking in the water or hiding under the car. Our only entertainment was an occasional visit by Ethiopian border guards who had a radio and could tell us the news.

This was the most boring study site I've ever had. We were happy to leave Lake Turkana, drive back to the highway, and continue south across red deserts with chimney-like termite mounds twenty feet tall. As soon as we got back on pavement, I tabulated the data. There was absolutely no difference between crocs in Lake Turkana and in the small ponds. Of course, it was only half of the answer. I still had to do a similar mini-study of alligators. But for now, I could get back to my most time-consuming chore: comparing crocodiles living in places with only big and only small lakes.

Sarit and Alex were traveling around Kenya at that time, and we planned to pick them up at the border, so that we could do the rest of our research in Africa together. It wasn't easy to arrange: we could communicate only by email and rarely had Internet access. When Tesfaye and I returned to the highway and got to a sizable town, I checked my email (it took three hours) and found that my friends had left Isiolo two days earlier.

Kenya consists of two very different parts. The southern part, the one that most tourists see, is mostly savanna, with big cities, excellent roads, and civilized-looking people. The northern part is desert, with scary-looking nomads, crappy towns, and some of the worst roads in East Africa. The main gateway between the two parts is a police checkpoint north of the town of Isiolo. Sarit and Alex were now making their way across the Dida Galgalu Desert of the North.

Tesfaye and I got to Moyale, the main border crossing to Kenya, and asked the locals if two white folks had recently crossed from the other side. A small crowd gathered, and boys ran to the border guards to ask them. Everybody agreed that there had been no foreigners and no buses from Isiolo for a week.

So we went to a café to eat. But before our injera was ready, something happened outside: there was running and shouting, then footsteps of a lot of people. Two kids burst into the room, screaming, "They are here! They are here!" Finally, a huge crowd escorted Sarit and Alex in. I felt like Stanley when he found Livingstone.

My friends were starving; they'd been on the road for three days. I told them how good Ethiopian food was, but when the injera arrived, it wasn't made of teff, and it had no stews, only salads and vegetable paste.

"Sorry," said Tesfaye. "The Lent. No meat for forty days. Maybe you get some when we reach Muslim places."

We drove back to the Rift Valley and followed the chain of lakes that stretches almost all the way to Addis Ababa. All of them had crocodiles, but there were too many people on their shores. Then we found a great place called Nechisar National Park. It was an isthmus separating two large lakes. The lakes had rocky shores with no ponds or marshes, and all the rivers flowing in were dry this time of year. There were lots of crocs, including hundreds of really big ones on a small peninsula called Crocodile Market. Observing them from tall lakeside cliffs was very easy. Interestingly, the presence of so many crocs didn't prevent the locals from angling for fish standing chest-deep in the water.

Few people come to Ethiopia to see wildlife. Its national parks have a lot of species unique to the country, but these species are little known to nonspecialists. There are no huge herds of big game, like in Kenya or other "safari" countries. As a result, these parks bring little revenue overall and virtually zero income to the local people. So the locals hate the parks. Poaching often turns into low-intensity guerrilla warfare, forcing the parks to require tourists to be accompanied by guides and armed guards. That doesn't add to their popularity among tourists—a vicious circle. It's a miracle that Ethiopian parks still have some amazing animals left.

I had a permit from Addis Ababa allowing us to travel and camp in parks without guides or guards, and we enjoyed the privilege.

We set up our tent near the car, and every morning a hornbill would wake us by pecking the car mirror with its huge bill, apparently trying to fight with its reflection. We'd watch the crocs for five hours, sleep through the scorching afternoon, then drive to a hot spring in the fig forest to soak in a natural Jacuzzi, and walk around for half the night looking for wildlife.

There were plenty of animals around, but our most memorable encounter was with a mouse the size of a small apricot we found one night. It was so tame that we could pick it up by hand and feed it pieces of cracker. Very small rodents in dry grasslands of other continents are also amazingly tame sometimes: silky pocket mice in the American Southwest, pygmy jerboas in Mongolia, harvest mice in the Russian Steppe. I have no idea why.

Nechisar crocodiles "sang" the same way as in all other places with big lakes or rivers: they simply headslapped or added a soft "cough" but never a loud roar. Now we had to find a place with only small lakes or streams. One old book mentioned crocs living in the deserts and swamps east of the Rift Valley, beyond the Bale Mountains. So we went there.

These were Oromo lands. The Oromo are one of the most ancient ethnic groups in the Horn of Africa and make up more than a third of the Ethiopian population, although they are seldom seen among high-ranking officials. Most are dirt-poor farmers or equally poor nomads raising camels and goats.

We tried looking for crocodiles in the desert in the far eastern part of the Oromia Region. But all streams were dry or almost dry and had no crocs. There were larger rivers near the Somali border, but driving there would take at least a week.

The only deep pools we could find were inside the beautiful Sof Omar Cave, where all crocs had been killed a long time ago. As a consolation, we found huge bat colonies in the cave, with a dozen different species. These were about to disappear as well, because a large team of workers was building a tourist trail and putting electric lighting in the cave. Some parts of the cave were very dry, and mummified bat corpses were hanging from the walls. A few mummies had little drops of dried blood in their nostrils, and I wondered if they'd been killed by Ebola virus or some other hemorrhagic fever. I later told a friend who worked for the Centers for Disease Control about these mummies, and now an international virologist team is planning to study that place.

Then we went to Harenna Forest, a small area of ancient rain forest that miraculously survived in the southern foothills of the Bale Mountains. It's one of five places in Africa where new kinds of larger animals are still being discovered. We saw a monkey that was later described as a new species, and a small black mongoose that was probably unknown to science.

Harenna is claimed to be the very location where coffee was first cultivated more than a thousand years ago. Local people prefer "wild" beans gathered in the forest and brew the best coffee I've ever tried (with Jamaica Blue Mountain Coffee a close second). When they invite you to visit their homes, they perform a complex coffee ceremony, almost as impressive as the tea ceremonies of Fujian or Japan.

The only river in the forest was too big for my study. We realized that we had to drive all the way to northern Ethiopia and look for a good study site there. To get back from Harenna, we had to go over the Bale Mountains; that was a problem because our brakes weren't working any more. We climbed to the beautiful, cold, wet meadows on the plateau, which looked a lot like Scottish moors. In the meadows there were all kinds of strange-looking birds and flowers, hares with ears larger than their bodies, and lots of large rats, hunted by coyote-like Ethiopian wolves that live only in the highest mountains. Driving down the steep grades in first gear took a full day, but in the evening we were rewarded with a free sauna at the headquarters of Bale Mountains National Park.

The forests around the headquarters had a lot of rare wildlife. The most interesting place proved to be the garbage bin outside the park office. I watched it for a few hours late at night and saw two striped jackals, one golden jackal, and a slightly larger doglike creature that I couldn't identify. Only two years later it became known that there's a tiny, isolated population of dwarf gray wolves in remote parts of Ethiopia.

We would have liked to stay there longer, rent horses, and explore the far reaches of the park—but there was a long, long drive ahead and more crocodiles to find.

27
Paradise in Hell
Crocodylus niloticus

Travel all the Earth and see how Allah began the Creation and how He makes a new Creation.
—Quran 29:20

We briefly stopped in Addis Ababa to exchange the car with faulty brakes for another one. The new jeep didn't have air conditioning, and the headlights barely worked, but we didn't have time to get it fixed. We drove north, through the central highlands of Ethiopia, home of the Amhara people.

The highlands were Tesfaye's homeland. He'd spent much of his life driving tourists around them and knew them well. But this time was different.

There are a lot of things you have to tell your driver in Africa. Some of them are difficult to explain. Because of high transportation costs, Ethiopia is an expensive country to travel, so it's mostly visited by tourists who don't care about spending a few extra dollars. We did. We had to camp or find the cheapest hotels in town, normally used only by locals. We cooked our own food every evening. We were on the move from dawn till late night. We taught Tesfaye to stop for all wild creatures on the road, even for the smallest snakes, so we could jump out and photograph them. I also explained to him that it was better to drive slowly through towns with lots of children and fast through the desert, not the other way around as African drivers usually do.

We passed Lake Tana, with its ancient island monasteries, and Gondar, the old imperial capital with an impressive castle and the most beautiful church in all of Africa. That little church looks a bit like the old pueblo churches of New Mexico, but inside are stunning seventeenth-century frescoes, painted on cloth glued to the walls with a mix of clay and sheep blood. On the ceiling are six hundred angel faces, outstandingly vivid and funny.

North of Gondar the road crossed the Simien Mountains. This was once the land of the Beta Israel, the Amhara people practicing Judaism. During the famine and subsequent civil war of 1985–1993, the entire tribe was airlifted to Israel. Pseudo-Jewish souvenirs are still sold to tourists on roadsides around Gondar, and thousands of Ethiopians demand the right to move to Israel claiming their Beta Israel descent. As for the mountains, they were declared a national park but didn't remain uninhabited for long. Christian farmers immediately moved in, so now only the highest summits retain natural vegetation.

These mountains have virtually vertical northern slopes, and the sweeping views from cliffs ten thousand feet high are sublime. Tourists, who have customs that are set just as firmly as those of Omo Valley tribes, traditionally go there for trekking. We didn't feel like hauling heavy backpacks just for the sake of it, so we simply drove to the highest part of the ridge to see rare walia ibex, huge lammergeier vultures, charming klipspringer antelopes, eagle-sized ravens, and, of course, the geladas. The Simiens have gelada baboons, the weirdest-looking monkeys in Africa. They look a bit like baboon-lion crosses, or, in the case of old males, like moving haystacks. They are very social and tame. In the morning, when they leave their sleeping places in the cliffs to graze in the meadows, you can walk straight in the middle of a herd. Geladas have the largest brains of all monkeys (excluding apes) and the largest herds, sometimes more than a hundred. Many zoologists think there's a direct relationship between brain size and herd size in every family of primates and often mention geladas as a good example.

Beyond the Simiens we followed an old road, built in 1938 by the Italians to deliver military supplies to occupied Addis Ababa. The road crossed a broad plateau dotted with rusty remains of Soviet tanks. During the 1980s, a widespread famine led to a massive insurgency against the Derg (the Communist government). The Soviet Union provided the Derg with an unlimited supply of weapons. The Derg used aerial bombardments of cities and tried to starve the rebels by blocking humanitarian aid to rebellious provinces. It didn't work. As soon as the Soviet Union began to disintegrate, the Derg fell, and the rebels entered Addis, using maps from the *Lonely Planet* guidebook to navigate the city streets. The current government, which at least pretends to be democratically elected, is dominated by the Tigray-Tigrinya, a northern ethnic group, so the far north of Ethiopia has by far the best roads and the happiest-looking people. The tanks are now cut for scrap metal, and the outer shells of unexploded bombs are converted into church bells.

We reached Axum, the oldest city in Ethiopia. It has been a capital of a large empire for more than a thousand years, starting from the fourth century BC. A church in Axum is one of the three places in Ethiopia claimed to contain the Ark of the Covenant. Nobody except the top-level clergy has ever seen this relic, but everybody is welcome to believe that it exists.

We were already close to the border with Eritrea, permanently closed after a bloody and senseless war in the 1990s. We hadn't seen any crocodiles since leaving the shores of Lake Tana, and the lake certainly didn't qualify for a "small pond" study site. We decided to return to Addis through northeastern Ethiopia, and, if we couldn't find crocs there, try Sudan the next year. We weren't happy about this. Sudanese national parks are infamous for $700-per-day entry fees.

As soon as we left the North, the roads got worse again. There was so much dust that Sarit and Alex had to wrap themselves in chitenges, the large pieces of cloth they'd bought in Zambia. It made them look like Mary and Joseph escaping to Egypt, with Sarit's huge camera lens also wrapped in cloth and looking like a baby. Local people, too, wrapped themselves and looked biblical, except for some young city girls who'd wear the traditional white sheets of cloth on top of tight jeans. But the most exotic pieces of clothing were huge umbrella-like hats made of press-dried dung. They were also used as lids for injera-baking stoves.

Despite all the dirt and dust, we liked Ethiopia more and more. We even got used to huge crowds of children following us through village streets, shouting, "Highland!" (the name of the first brand of bottled water in Ethiopia, now meaning bottled water in general), "Birr!" (Ethiopian currency), "Give me!" or "You!" It's widely believed in Ethiopia that constantly shouting, "You!" is a polite and friendly way of addressing a foreigner. For the kids it's just a game, and if you switch their attention to something more interesting, such as talking or being photographed, they instantly become normal inquisitive children.

Women would often encourage small children to run up to us and hug our legs, asking to be picked up. Tesfaye explained that these women hoped that their kids would get adopted and taken to the West. This was surprising. Just a year earlier we were in Central America, where village folks were often paranoid about evil gringos kidnapping their children for adoption.

The only people we didn't like were priests and monks, and there were lots of them. Every village had at least one big church, usually on a hilltop. There were also a few mosques, but these were located near water, in deep river valleys, and not visible from the road.

We didn't care much for standard-looking modern churches. Being isolated from the rest of the world by hostile Muslim peoples, the Ethiopian Orthodox culture went into a decline in the Middle Ages and never recovered. But there were a few ancient ones, some dating back to the fifth century, that we really wanted to see. We had a letter from a very high office in Addis granting us free entry to all churches, monasteries, and mosques. But the gatekeepers ignored it, even if Tesfaye read it to them (many monks and priests were illiterate). They demanded exorbitant sums for letting us in, especially if the church was difficult to reach (accessing some rock-hewn churches required almost-technical rock climbing). The gatekeepers were the only people in Africa who didn't succumb to Sarit's persuasion. Even Tesfaye got mad at them, and frequently quoted the Bible verses about Jesus casting out merchants from the Temple.

But that wasn't all. Sarit sometimes had to wait for us outside, because many monasteries didn't allow women (or even female cattle) on their grounds. And every time she sat down anywhere within a mile of a church or a monastery, she was immediately surrounded by junior clergy who tried to sneak a peek under her chitenge.

Finally, we could leave the cool highlands for the hot deserts to the east, inhabited only by Muslims. We decided to make a little side trip to see the most unusual part of the country.

In far northeastern Ethiopia, near the borders with Eritrea and Djibouti, lies the Afar Desert. The word *afar* means "ash" or "dust" in Semitic languages. This is probably the place where humans evolved from apes about five million years ago; the most sensational transitional fossils have recently been found here. At that time the region was covered with trees, but now there's little vegetation. There are major junctions of continental rifts in Afar, and the Earth's crust is constantly breaking up, producing volcanoes, lava flows, and deep fissures. The northern part of Afar, called Danakil, is the lowest place in Africa, almost five hundred feet below sea level. There are no weather stations in Danakil, but it's believed to be the hottest place on the planet, with daytime temperatures seldom falling below 120° F.

The desert is inhabited by the Afar people, who are usually described as warlike, semi-wild brigands who make their living trading salt from dry lakes and robbing caravans. They used to have a custom of castrating uninvited visitors, and even today travelers through their lands occasionally get kidnapped. I knew that peoples with such reputations tend to be the most interesting and pleasant to meet.

By the time we descended from the highlands it was dark. In the first Afar village we got a bit lost among countless vehicle tracks in the dry riverbed, and stopped to ask directions from a group of girls walking along. We were instantly invited to a house, fed delightful hot flatbreads with acacia seed, given excellent coffee, and offered a room with walls woven of dry branches. All night we could hear the soft steps of camel caravans, walking along the village's only street in the moonlight.

Our hostess's name was Eisha. In the morning we got out an inflatable globe we had for such occasions, and showed her where we were from and what our route was. Sarit was given a lot of bead jewelry and exchanged her Zambian chitenge for an Afar-made one. Seeing her walk around in a local cloth drove people crazy with excitement.

Afar has horrible roads, winds so hot they can burn your face to blisters, and boring landscapes. The few people who visit it arrive in large jeep caravans and pay about $2,000 for a five-day trip. Traveling in one car is not only dangerous but also almost impossible. Every time you pass through another clan's territory, you are required to take up to ten guides and guards with you. In addition to paying for trespassing, you have to pay all these people small wages, feed them, and provide water and blankets. We tried to resist the best we could, making friends with every sheikh and army commander along the way. We ended up having only two extra people most of the time, although in one particularly dangerous place we had to carry six soldiers and a machine gun on our roof. Cooking for six, not to mention twelve, on our tiny camp stove took a lot of time. The car was so crammed that we could take little water, and if we had ever got stuck or the car had broken down, walking out would've been a survival test.

But we'd never trade that journey for an organized trip in a luxury air-conditioned jeep. We met the most hospitable people we've seen in Africa. Men were invariably armed to their file-sharpened teeth. Girls were stunningly beautiful; they walked around naked above the waist, but with thin veils over their faces. We were invited to every tent we passed, and leaving was always difficult. We even learned to play the Afar version of *mancala*, an ancient African board game, played in the sand with dry goat feces for pieces.

Danakil has two places worth seeing. In the far north, at the Eritrean border, lies an almost-dry Lake Assal. Every day four thousand camels arrive there, caravan after caravan, to be loaded with salt. In the middle of the lake there's a colossal bubble of rock salt, covering a volcano called Dallol. The hot springs at the summit are by far the weirdest springs anywhere on Earth. There are blood-red terraces, smoking white domes, and large pools of green, steaming sulfuric acid with rims of bright-yellow brimstone.

Fifty miles south, lost among the endless dust deserts, is another volcano, called Erta Ale. It has a lake of boiling lava in its crater. At that time there were only four such lakes in the world, all difficult to get to. I'd seen the lava lake in the crater of Mount Nyiragongo Volcano in Central Africa a few years earlier, but that crater was very deep. The lake in Erta Ale crater is smaller, but you can watch it from fifty feet above. Sarit and Alex had never seen one, so they trusted me when I told them it was worth *any* effort.

Climbing Erta Ale over black lava fields would be easy if it weren't so hot there. Tourists arriving in air-conditioned cars often die before getting to the summit. We spent a night at the edge of the lake, watching it churn, boil, constantly create huge bubbles, small explosions, and beautiful patterns of black and red. At sunrise, just as we were about to leave, the entire surface of the lake exploded, sending huge "drops" of red lava three hundred feet into the air. By the time these drops fell back, they were already black rocks.

The return drive was difficult. It was a very hot day, probably about 125° F. We had a strong tailwind, so the radiator didn't work well, and we had to stop every few minutes to cool the engine. But we couldn't open the windows because we were driving through a dust storm the whole time and the car would instantly fill up with dust. We had half a quart of water left. We soaked a cloth and wrapped it around Sarit's head, but it was dry again in two minutes. Our armed guard, a young kid, was so terrified that we had to let him out of the car at every stop. Our guide, an old Afar warrior, lost consciousness, and he was the only one who more or less knew the way. Then Tesfaye also collapsed, and Alex looked like he was about to. I navigated by the shape of low dunes and the direction of the wind. For the first time in all our travels together, I was afraid that I might lose my friends. But in the afternoon the wind suddenly changed, the dust settled, I rolled down the windows and floored the gas pedal. Within five minutes everybody was coming back to life. Two more hours of shaking in the rocky foothills, inhabited only by incredibly hardy gazelles and bustards, and we entered a deep canyon. There we found water.

Saadi, a great thirteenth-century Persian poet, once said: "If you want to experience absolute pleasure, go to the desert for two days.

When you come back, the first muddy puddle you find will give you absolute pleasure." But this wasn't a muddy puddle. It was a tiny crystal clear waterfall with a small lake beneath, the only fresh water for fifty miles around. There were even frogs in the lake. It was paradise.

The next day we returned to the district town and said good-bye to our Afar guide. We were about to leave when a pretty girl ran up to us, shouting excitedly. "She says," translated Tesfaye, "that Sarit is wearing her sister's cloth." The girl was Eisha's sister. She had come to town to apply for a job in the local tourism office. We walked with her to the office, showed all our credentials one more time, and told them that her family had been very helpful to foreign visitors. She was immediately hired.

We gave her a ride back to Eisha's village, where we discovered that Eisha had become a local celebrity. She was the only person in the district to have ever had foreign guests. We charged our laptop from the car battery and spent the evening showing photos from other parts of Africa to the entire village.

In the morning we drove up to the highlands, to the cool air, green trees, and paved roads. And suddenly we all felt that we were dying to return to Danakil: to that unique place where Muslim prayers are given not facing Mecca but with your back to the wind; where bottles are opened with AK-47 magazines; where plastic fuel cans are kept as family heirlooms but women wear five-ounce golden earrings. To the world of slender people, slim camels, and upbeat larks; the universe of black lava, yellow sulfur, white salt, and gray dunes. To the land that had been our home for five days.

Our adventures in Afar weren't over yet. We had to cross its more developed southern corner using a segment of the paved highway leading to Djibouti. That highway was built during the war with Eritrea, when landlocked Ethiopia lost access to the coast. The new road was the country's lifeline, so there was heavy truck traffic.

It was a sad journey. The local Afars had been civilized. The men had mostly left to work in cities; the women were standing along the road with empty plastic bottles, begging truck drivers to share some water. Nobody ever stopped.

The road crossed a national park called Yangudi Rassa. It had been created to protect one of Africa's rarest animals, called an African wild ass. These ancestors of domestic donkeys are beautiful desert animals, but they survive only in a few remote corners of Ethiopia, Eritrea, and Somaliland. In the early 1990s I participated in a breeding project in Israel, trying to get them to reproduce in captivity. I really wanted to see them in the wild, so we spent a couple hours driving around the park. There was plenty of cattle everywhere but almost no wild animals. Once we saw a possible wild ass, but it was too far to be sure.

We drove to the next town and found the park office, a tiny plywood shed occupied by two park employees and their families. One of them spoke a little English.

"Do you have any wild donkeys in the park?" I asked.

"No."

"What do you have there?"

"Problems."

Well, at least he was honest.

We decided to look for crocodiles in desert ponds around the city of Harar in far eastern Ethiopia, near the border with Somaliland. At the edge of the Afar Desert we camped by a small stream. The next morning Sarit woke up with acute stomach pain.

We drove toward Harar, hoping that she would get better. But she didn't. The road was long and slow, snaking along the top of a mountain range separating Afar and Ogaden Deserts. Sarit was almost screaming with pain. There was a small town called Mieso along the way, and we found a hospital there.

Having heard all kinds of horror stories about African hospitals, we expected to see a stinky cholera ward, with the dead and the still living lying in heaps on the dirty floor, the only ventilation being provided by the wings of myriads of flies, syringes being sterilized with urine after every thousand injections, and clumps of Ebola and Marburg fever viruses dropping from the moldy ceiling. But the hospital, although a bit spartan, was reasonably clean and effective. Within an hour they ran a few tests, diagnosed Sarit with acute gastritis, gave her an injection, and prescribed some pills that we bought in a pharmacy across the street. By late afternoon Sarit was merry as a little bird again. It all cost us less than $7. The next time I get sick, I'll consider flying to Ethiopia.

Harar is the unofficial capital of Ethiopian Muslims. It's believed that Prophet Muhammad had once sent his family here to seek shelter during the times of persecution. The natives of this semiautonomous city speak their own language, called Adare. Like most ancient Muslim cities around the world, it has narrow streets, dilapidated buildings, huge piles of garbage, and insufficient water supply. Locals believe that foreigners have no right to walk the streets without a guide. If you ask them to leave you alone, they tell you it's a free country and they can walk wherever they want, and then keep following you, loudly cursing the heartless palefaces. Instead of "You!" Harar people scream "Farenji!" when they see you. The word dates back to the Crusades and used to mean "a Frank" in Arabic, but now it means "a Westerner" throughout the Islamic World and beyond.

There's so much garbage in Harar that goats, dogs, rats, and vultures can't recycle it all. So the city is inhabited by a few hundred spotted hyenas. They are well behaved, show up in the streets after dark, and only occasionally supplement their diet with a child or some part of an adult. About forty years ago one local man started feeding them, and nightly hyena feeding became a tourist attraction. You can hand-feed and even carefully pet them. Spotted hyenas are really cool animals, but they aren't very popular, judging by the fact that very few tourists ever visit Harar.

In fact, few tourists ever show up anywhere in eastern Ethiopia. After we failed to find crocodiles around Harar, we decided to try Awash National Park on our way back to Addis, and were surprised to be the only visitors there. It was a wonderful place, with lots of antelopes, birds, and rarely seen hamadryas baboons, the world's only desert-adapted primates. It also had crocodiles.

We finally found a perfect "small pond" site. Awash River flows from the highlands to the desert through a deep canyon. Rockslides have broken its bed into a series of pools separated by shallow rapids, and waterfalls, and each pool had a few crocs. We recorded quite a few "songs" there. Then we moved to an even more scenic place a few miles away that also had crocodiles. Under a black lava cliff ninety feet high there were five small lakes fed by crystal-clear streams with warm water. Up the streams we found hot springs, turquoise pools hidden in a palm grove. The water in the pools was almost too hot to sit in, so we'd go there only at night, when the air was slightly cooler.

Awash crocs roared almost every time they "sang," and their roars were loud enough to be heard from at least three hundred yards away. My theory still held.

We got the necessary number of records and drove to Addis. We could tell that the capital wasn't too far, because we started seeing people selling wild birds in small cages along the road. Tiny, emerald-green Abyssinian lovebirds were the most popular. Selling wild birds was illegal, so every time we would stop, show these people our papers from Addis (they couldn't read them anyway), tell them that next time we'd arrest them and put them in jail, confiscate the birds, and release them in the next clump of trees.

Tesfaye was happy to get back in one piece. He bought such a huge bag of teff grain for his family that our car almost fell apart when we loaded it on the roof. After watching us bargain over every birr for two months, he clearly didn't expect a tip, but we decided he deserved a good one for all his suffering. He later wrote me that it was his best journey ever. But, despite all my lectures on geology and astronomy, he still felt obliged to believe that the world was only a few thousand years old.

28

The Land of Lost Opportunities

Osteolaemus tetraspis

If a road is built through your garden, don't expect to see your fruit ripen.
—Chinese proverb

THE NILE CROCODILE AND ITS RECENTLY DISCOVERED RELATIVE, the sacred crocodile, aren't the only crocs in Africa. In the forests of West and Central Africa live other species, much less known. Just about all forest fauna of that region is poorly known. For many decades zoologists have worked mostly in the savanna parks of East and Southern Africa. A few forest animals, like the aquatic genet and the owl-faced monkey, have never been seen alive in the wild by a scientist.

It's not that the naturalists have been lazy. Of course, studying animals in dense forests is more difficult than in the open savanna, especially since there's more hunting in the forests than in the well-protected parks of the East and the South. But the main problem is getting there and back, and not being killed in between. Most countries of the forest zone are former French (or, in one case, Belgian) colonies, so they have excruciatingly complex, corrupt, and inefficient bureaucracies. Despite being very rich in natural resources, they are prone to frequent armed conflicts, coups d'état, debilitating poverty, and major infrastructure problems, This is especially true in the Democratic Republic of the Congo, formerly known as Zaire, which has endured the most brutal colonial regime in history, followed by forty years of civil war. Congo-Zaire was the African country I wanted to explore the most, for it has the world's highest concentration of mysterious, poorly known animals.

During my first trip to Africa, in 2005, I spent a few days there and had more adventures than in the remaining four months of the journey. I climbed to a lava lake and then talked a missionary into flying his small airplane to the remote Ituri Rain Forest to look for okapis. He couldn't pilot the plane himself, and for me it was my third flight without an instructor. We landed on a sandbank of a forest river and lived among Pygmies for three days, then flew over the famous Garamba National Park on the Sudanese border. I got to see an okapi and lots of other fauna, including one of the world's last northern white rhinos (a year later they were hunted to extinction in the wild). The flight wasn't all fun. We had to taste human meat during one of our landings (refusing it would offend our hosts). Then we accidentally flew into Sudan, barely escaped being shot down, ran out of fuel, and had to land on a military airfield in Uganda, where we almost got arrested. The missionary promised to have me killed if I ever set foot in Congo-Zaire again, but I thought he should've been grateful that it didn't end worse.

Unfortunately, any serious trip to Congo-Zaire requires a lot of time and money, and we had neither. So we decided to go to the People's Republic of the Congo, which is a bit better off, despite a recent civil war and a few armed gangs on the loose.

Congo was the only country I've been to where officials at the airport didn't know the word *tourist*—we had to explain it to them. The government website claimed that you could get a visa on arrival, but when we landed in Brazzaville, the capital, we found that it was only possible if you had an invitation and a hotel reservation. We had to hire a local guy to take a taxi to a hotel and get us a reservation. He later asked fifty bucks for the favor.

Every time you arrive in a new country, it seems expensive, but it gets cheaper as you learn to follow local ways and avoid local scams. Congo seemed outrageously expensive. We felt that something was

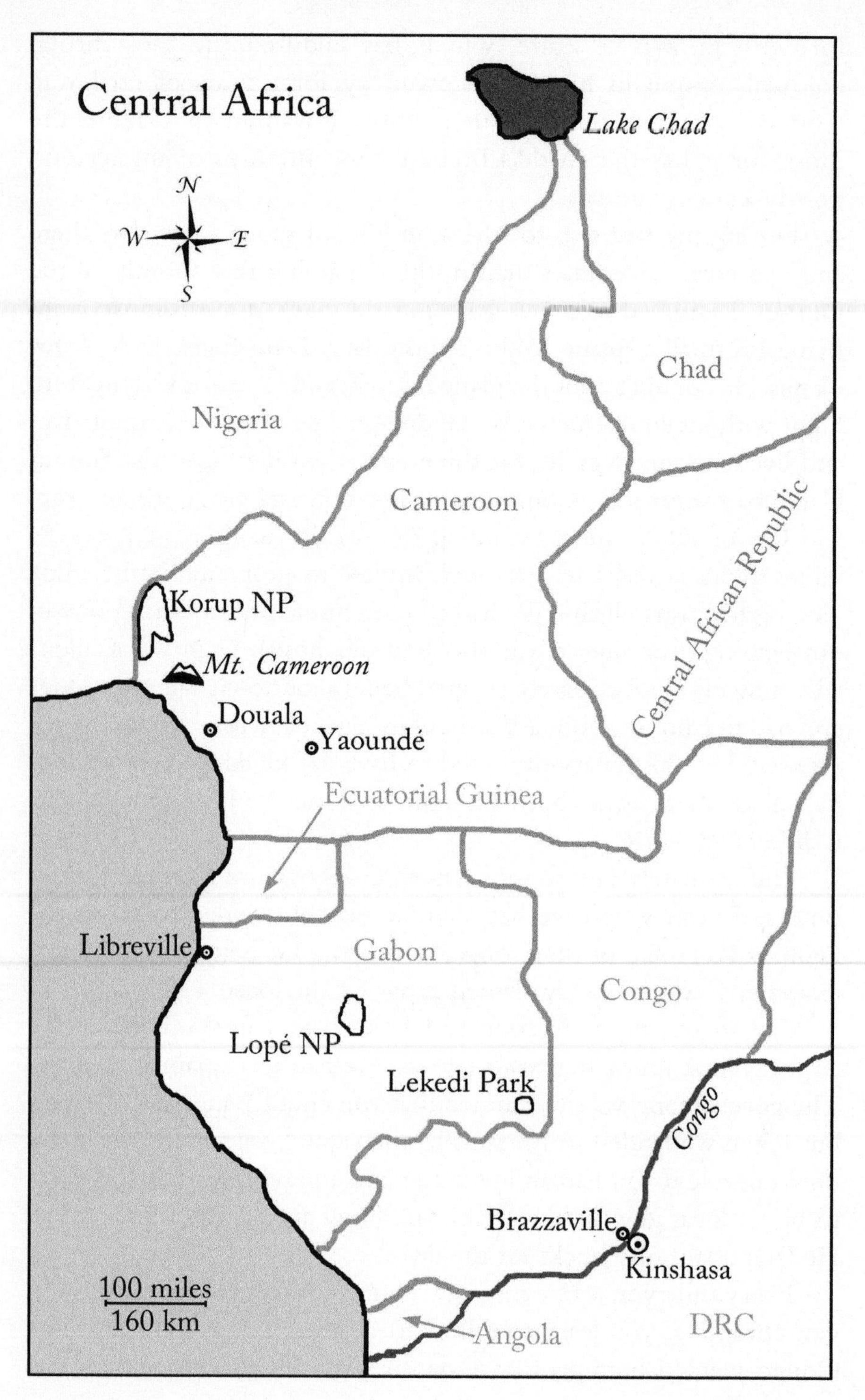

Central Africa
Lake Chad
N
W
E
S
Chad
Nigeria
Cameroon
Central African Republic
Korup NP
Mt. Cameroon
Douala
Yaoundé
Ecuatorial Guinea
Libreville
Gabon
Congo
Lopé NP
Lekedi Park
Congo
Brazzaville
Kinshasa
100 miles
160 km
Angola
DRC

wrong, that a country so poor just had to be cheaper. But for the moment everything—hotels, taxis, restaurants—seemed so overpriced that it would drain our budget before we could find cheap food, shelter, and transportation. The fact that my French was pretty basic, and my friends didn't speak any at all, didn't help.

But we got lucky. We met a local man named Jean-Claude. He looked like an African version of Bilbo the Hobbit. After just a short conversation he invited us to stay at his place. A typical house of a middle-class Brazzaville resident, it was a concrete cube with a small sandy yard and a tall metal fence, hidden in a narrow street covered with a thick layer of plastic debris. Although the city was on the shore of the world's second largest river, the mighty Congo, there was no water supply to the house. The electricity was on for only a few hours a week.

The Congolese were very different from the people of East and Southern Africa. Nobody shouted, "Hello!" to us in the street or begged for money. We were probably the only Westerners to have ever set foot in that part of town, but we could walk around as if unnoticed. The people were calm, polite, and friendly, and it was very unusual for a conversation to go on for more than two minutes without everybody laughing. If, slowly rolling down a busy street in a taxi with its windows down, you sneezed, a few passersby would wish you good health.

Here the rainy season had just ended, so it was very hot and humid. Even the locals found this oppressive heat stressful and carried large cloth sheets for wiping off sweat. Unfortunately, French missionaries have doomed the people of Congo to eternal suffering by teaching them that it is improper to wear short pants and sleeves. We, too, had to follow these stupid rules. If I ever become a local tribal chief, my first decree will ban all clothing, and the second will expel all missionaries.

Our stay in Brazzaville was a naturalist's nightmare. Just two hundred miles to the north were endless rain forests full of wonderful animals, but we were stuck in the city. Everybody was very helpful, but the process of getting permits was so long and complex that it took more than a week. We spent most of that time in a downtown restaurant that had air-conditioning and free Wi-Fi. It was too expensive to buy anything but ice cream there, but we could get French bread and Turkish shawarma in the street. Before going back to Jean-Claude's place, we would buy dinner for his family. There were two supermarkets, one of them French, where the food and prices reminded us of Paris, the other Chinese, where everything was like in Beijing. Later we learned to buy food in a street a couple blocks from Jean-Claude's house, where it was fresh and the prices were African.

After a week of waiting we were so fed up with eating ice cream and sleeping in a hot room with no windows that we decided to make a short trip across the river. On the other side, just a mile away, was Kinshasa, the capital of Congo-Zaire. It took us two days to get a visa, thirty minutes to cross the river, and four hours to fill all the forms at the passport control. Jean-Claude, who was a member of the local Evangelical Church, called his friend Jerome, who was a member of a similar club on the other side, and arranged for us to stay at Jerome's place if we got stuck in the city. Jerome had a rare luxury, a cold shower, which we enjoyed more than any shower in our lives.

Our goal was to reach a small nature reserve outside Kinshasa, the only remaining forest in the area. When we got there, we learned that there were no crocodiles left in it. But the reserve had some bonobos.

Bonobos live only in Congo-Zaire, inside a huge loop of the Congo River. They look like a smaller, more delicate version of chimps but are much more humanlike in behavior. In popular literature they are sometimes called "hippie chimps," because their society is based less on violence and more on sex. They don't have intertribal warfare or infanticide but practice heterosexual and homosexual sex in more than twenty positions, and do so very frequently. We managed to observe them in the forest for about ten minutes (which is more than most zoologists ever get), and even in that short time we saw the bonobos do things that chimps never do. A female walked upright all the time, carrying a baby at her breast exactly like a woman would. A male waded waist-deep into a lake and washed himself (chimps are afraid of water).

In both Brazzaville and Kinshasa, people think that the other city is dangerous, full of gangsters and corrupt policemen. But we never had any problems in the streets on either side of the river. Everything went fine until we tried to get on a boat back to Kinshasa. The border officials threatened to arrest Jerome for unauthorized communication with foreigners, and we had to buy him out. That two-day trip to Kinshasa cost us $300 per person; almost all of that money was spent at border checkpoints.

We finally had the coveted permits to visit a national park, and all we had to do was find a jeep. Rental prices were insane. Jean-Claude, who worked as a driver for a Chinese company, introduced us to his boss, Mr. Hu. We talked Mr. Hu into going to the park with us to see gorillas. We would return to Brazzaville by hitchhiking; it's always easier to hitchhike from a small place to a large city than in the opposite direction. We couldn't wait to get out of Brazzaville.

The next morning Jean-Claude took us to Mr. Hu's office, went to get their 4 x 4 truck from the garage, and disappeared. Five hours later we learned that all the lug nuts from the wheels of the truck had been stolen and would take weeks to replace. Our visas were to expire in ten days. Even if we did somehow manage to get to the forest, there would be no time left to study crocodiles. All the time and money we had spent in Congo were wasted. We had no choice but to try to get to Gabon. Another day was spent getting Gabonese visas, buying farewell gifts for Jean-Claude's family, and figuring out how to get out of Brazzaville.

We took a bus to central Congo. It took the bus crew four hours to sort out which passengers and whose luggage to take on board, but then they took off at really scary speed. We followed the country's only paved road, narrow and winding. The pavement ended in the native city of the country's president. From there we had to hitchhike.

Getting to the border took us three days because there was almost no traffic. Sometimes we would get a lift, but the car or minivan would soon get stuck in the mud. We passed five checkpoints, every time spending an hour to let the soldiers write down our passport data letter by letter. The drivers also had to pay bribes.

But the journey wasn't unpleasant. It was no longer hot and we crossed lush green savanna with patches of forest. There were few villages along the way, but we saw almost no wild mammals and few birds.

Some of these birds were unique to that part of Congo, and I think a hard-core bird-watcher would die to catch a glimpse of them. In the villages we could buy sweet lemonlike fruit, canned sardines, and live lungfish. If we stopped to eat, the entire village would gather to watch us. At first everybody would run away screaming every time we got a camera out. But sooner or later some kid would master enough courage to have his photo taken. Other people would look at his picture on the camera display, discuss it merrily, and line up for their turn. Digital cameras have made travel in remote areas so much fun!

I kept checking every river and pond for signs of crocodiles but didn't see any. Then we found a dwarf crocodile about three feet long, but it had already been caught. A local woman was transporting it to a town market with its jaws and legs tied up.

These cute "toy" crocs seldom grow over six feet long. They are the most common crocodiles in the forests of West and Central Africa, where they live in small lakes, ponds, and swamps. After the larger species were hunted out in many parts of their range, the "dwarfies" started colonizing mangroves and savanna lakes.

Just half a century ago you could find dwarf crocodiles in every puddle within the rain forest zone. But then the so-called "bushmeat trade," commercial hunting of wild animals for sale in city markets and restaurants, began to grow rapidly. As more logging roads penetrated the wilderness, the bushmeat trade turned into a scourge of African fauna and emptied the forests, affecting even the remotest areas. Primates, antelopes, porcupines, wild pigs, pangolins, and forest buffaloes are the most sought-out species, but everything from tortoises to pigeons goes into the pot. Dwarf crocodiles (as well as turtles, monitor lizards, and pythons) are particularly well suited for use in the bushmeat trade because, once caught, they can remain alive for a long time. So they are trapped relentlessly and have become rare and shy. I later realized that I had probably missed quite a few before I figured out just how difficult they were to see.

We expected the road to keep getting worse as we were nearing the border, but at some point it began to improve. There was a lot of road construction, with crews of burly Congolese workers invariably supervised by young, thin Chinese foremen, and in one case by a petite but obviously very brave forewoman. Sometimes the Chinese were so pleased when I used my small collection of Mandarin words that they ordered their tractor drivers to transport us for a few miles down the road.

You meet Chinese engineers everywhere in Africa. China is investing in African infrastructure; in exchange it gets concessions for extracting mineral resources made accessible by this infrastructure. The engineers mostly come from inland China, not from the coast, where it's easier to find a good job. Many of them are almost kids: Mr. Hu, who managed a company about to provide all of Brazzaville with a reliable electricity supply, was twenty-four. They usually speak little English and almost no French. They supervise dozens or hundreds of local workers and are driven to the edge of insanity by the local laziness, indifference, and lack of intellectual development. They hate Africa and pay their cooks a lot of money for Chinese dishes. They count days until the next annual vacation, when they can fly home for one week. With very few exceptions, they are totally deprived of the company of Chinese women, and African women don't take them seriously.

In the last ten years they've built and paved thousands of miles of road in Africa. Where they haven't gone yet, local roads are usually deep ruts in bottomless mud. They install power lines and cell phone towers, bridges and factories. Any construction takes at least ten times longer in Africa than in China, but they don't give up. They battle the world's stupidest bureaucracies, most corrupt governments, worst diseases, and the overall chaos of dysfunctional African life.

The Chinese foremen consider themselves the forerunners of civilization in the wilderness. They are always helpful, polite, clean-shaven, and reasonably clean, even at the dustiest construction sites. They are very democratic with their Chinese subordinates and don't mind playing a card game with them during the fifteen-minute lunch break. But they always keep a certain distance from Africans. Perhaps they have the right to do so. They are bringing Africa a future without river blindness, guinea worm, sleeping sickness, and famine—a future with the Internet, refrigerators, and good hospitals. They will modernize the life of the locals no matter how much the locals resist it.

They have never read Kipling, but perhaps they should.

Eventually, we flagged down a truck with Muslim drivers from Nigeria. Surprised by our "As-salaam alaikum," they let us ride in their empty trunk all the way to the border, where we spent the few remaining hours of the night on the floor of the Gabonese border checkpoint.

Gabon had already been thoroughly worked over by the Chinese. The highways were excellent, the ancient French-built railroads had been repaired, and passenger trains were running on time. People didn't know what a power outage was. The Internet was lightning-fast even in small towns. All villages, even remote forest outposts, had cell phone coverage. Everybody looked healthy and happy. But come to think of it, most people in Africa look happy no matter what.

Of course, there were still problems. The government was still corrupt, people said. By their estimates, in Gabon the government pocketed about half of all revenue, compared with three-quarters of it in Congo and pretty much everything in Congo-Zaire. The bureaucracy was still French-style, which meant nothing was ever easy or certain. There was still some deforestation and lots of bushmeat trade, although in Gabon illegally hunted wildlife was going mostly to expensive restaurants in the capital rather than to small town markets.

Gabon has an excellent network of national parks, the best in Africa's rain forest zone. If there was a place in Central Africa where we could find forest crocodiles, it was Gabon.

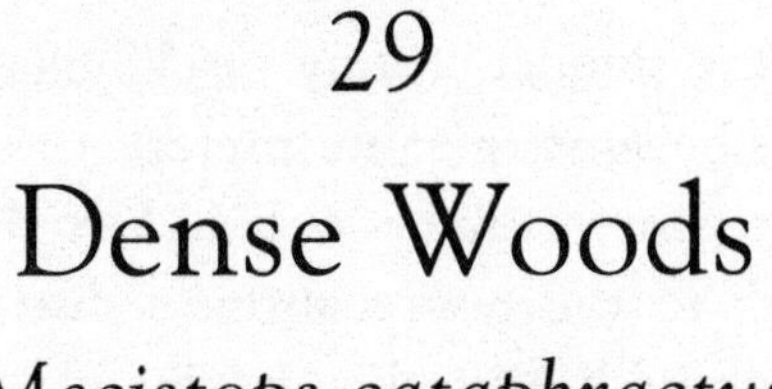

29
Dense Woods
Mecistops cataphractus

He who does not travel does not know the value of women and men.
—Berber proverb

In the far southeast of Gabon there is a large manganese deposit. A mining company used to transport the ore to the Congo River via the world's longest cableway, a fifty-mile stretch. The cableway went over rain forest and would have made a top-class tourist attraction, but it rusted away after the company switched to exporting the ore by railway to Libreville, Gabon's capital. A small town where cableway workers lived was suddenly left without jobs. To provide its people with income, the company created a beautiful nature reserve called Lekedi Park. It attracts over a thousand visitors every year—a huge number by Central African standards. We stopped there for a week.

The reserve was managed by a French gentleman who lived there with his wife and two kids. He hired a private tutor for the kids, a young soft-spoken guy named François. After graduating from college, François was looking for a job, but instead of a good salary and a promising career he wanted to get to a place as remote and exotic as possible. So he answered a newspaper ad and—voilà!—got a dream job in Lekedi. He could go into the forest with visitors, enjoy python meat in the town restaurant, watch giant moths flutter around the streetlights, and court slender, ebony-black village girls. When we showed up, François became our interpreter.

We met two other interesting people in Lekedi. One day we tried to hitch a ride back to the reserve from a district town and were amazed to see a truck with Australian license plates. The driver was a tall man named Mike who looked and sounded a lot like Paul Hogan, the actor who had played Crocodile Dundee. He told us that he'd had his old Toyota delivered by sea from Darwin to Kolkata and had driven it to Gabon through the Middle East, Europe, and West Africa. His plan was to get to Cape Town via both Congos and Angola.

On the same day a lady with a huge backpack arrived at the reserve rest house. Fifty-something, Giaconda was from Sicily. For six years she had been backpacking around the Sahel (the southern borderlands of the Sahara), studying the vanishing cultures of oasis fishermen.

In the evening, when we were all sitting around the campfire, our conversation somehow jumped from African to Middle Eastern politics, and François mentioned that he was Jewish. Then we discovered that we all were Jewish by origin, although none of us had a religion. Everybody was surprised, except me. I noticed a long time ago that Jews are disproportionately numerous among crazy people traveling to the remotest parts of the world for the strangest reasons. In back-country Mexico, you have to go only a hundred miles south of the US border to start seeing more Israeli tourists than North Americans. I think it has something to do with genetics. After all, the word *Hebrew* probably comes from the Proto-Semitic *habiru*, "wanderer." Perhaps the gene variant $DRD_{4\text{-}7R}$, associated with curiosity and restlessness, is more frequent in Jews? I've never considered myself belonging to Jewish culture (there's not much in it for an atheist), but I always feel myself strangely at home when visiting nomadic people, be it in Mongolia, the Navajo Nation, the Sinai Desert, the Siberian tundra or Afar. I guess I should be grateful to my ancestors for providing me with a gene that has made my life so interesting.

Lekedi Park is full of wildlife, but its forests are dense. At night there's a lot of small stuff to see up close, like wonderfully bizarre katydids, colorful tree frogs, and bright-eyed bush babies, but during the day you either catch glimpses of animals sneaking behind the trees or see monkeys and birds in the canopy more than a hundred feet above ground. Since almost all Central African animals are little known, even brief sightings are sometimes really exciting.

Probably the most beautiful creatures in Central African forests are guenons. There are over twenty species of these small monkeys in Africa; some of them are among the brightest-colored mammals, with green, red, orange, blue, black, and white fur. But these colors are mostly on their faces, probably helping to avoid hybridization between closely related species. Seeing guenons well enough to identify them is difficult, so I didn't believe my eyes when I saw a group of guenons with bright orange tails that I'd never seen pictured in books. Later I learned that they belonged to a species discovered only the previous year. Guenons can make amazing leaps from tree to tree because they are incredibly light: a guenon sitting on the outstretched palm of your hand feels like a pigeon.

There were also large meadows in the reserve, good places to see red river hogs, bright-red forest buffaloes, orange sitatunga antelopes, and the only mammals that are colored brighter than guenons, the amazing mandrills with their blue-and-pink faces and butts. There were even leopard tracks on the trails, but we never managed to see any cats.

I checked a few forest ponds in Lekedi, and found one dwarf crocodile. I watched it for a long time, but it didn't "sing." I asked François to question park rangers, and they all said that the time was wrong. Lekedi crocs "sang" from September to November, at the end of the dry season. We were off by six months.

Our only chance to catch the crocs "singing" was to move from the Southern Hemisphere to the Northern, where the seasons would be reversed. Fortunately, we were just 140 miles from the Equator.

A short train ride got us to Lopé, a large national park in the northern hemisphere part of Gabon. We talked to the park personnel, and they agreed to let us stay at a tourist camp deep inside the park. But just before we got a ride there, the country's minister of forests and tourism arrived for a visit with a huge retinue. He was an old man, very fat, with an air of self-importance. We knew from talking to park employees and local environmentalists that he wasn't a particularly helpful or friendly person, and there were also rumors of massive corruption. He saw us walking around, asked who we were, and said:

"I went to your embassy some years ago and asked them to invite me to visit Russia. They said I'd have to pay for the flight. They also said it would take two months to arrange. I will not be treated like this. You don't have my permission to be in Gabon. Go to your embassy and tell them to send me an official request for help. Maybe if they ask me, I will change my mind. I don't like Russians, I don't like journalists, and I don't like you."

The Russian embassy was two days away. Of the three of us, only I had Russian citizenship. Besides, Russian embassies are usually not the most helpful ones. We had to change countries again.

We hitchhiked toward the border with Cameroon and along the way camped near a large, beautifully forested river. Early in the morning I walked along the sandy beach, watching newly hatched soft-shell turtles the size of an Oreo cookie as they emerged from the sand and ran to the water. Some of them could eventually grow to over six feet. Farther ahead a small creek was entering the river. I followed it upstream, where its bed became rocky, with a series of murky pools. Then I saw a crocodile ahead. It was yellowish, with dark caiman-like spots, and had big, sensual eyes. It also had a long, narrow snout, almost like a gharial, so I knew it was the aptly named narrow-snouted crocodile.

This species has approximately the same range as dwarf crocodiles and is believed to be related to them, although they look totally different (dwarf crocs have short, broad snouts and are very dark-colored). It is known to be particularly good at tree-climbing, sometimes basking in the canopy of the shady forests it lives in (American and Johnston's crocs are also proficient climbers). It inhabits overgrown rivers, but females often move to densely vegetated small streams and flooded forests before nesting. I knew from experiments that the sounds of headslaps quickly lose their sharpness when spreading through water with lots of obstacles, so I expected narrow-snouted crocodiles to use both headslaps and roars. There were no published observations of their "songs," so I really wanted to study this species. But the crocodile I found jumped into the water as soon as it noticed me, and I could see by the trail of churning mud it left in its wake that it was moving underwater toward the big river. Besides, it was probably too young to "sing." It was the only narrow-snouted crocodile I ever saw in Africa.

Later I observed a pair of adult crocs of this species in a zoo in Florida. They had small babies and were very protective of them, but the male would still head-slap and roar a lot. People working with narrow-snouted crocodiles in other zoos have also told me that their crocs were unusually active "singers," roaring and headslapping many times a day. I still hope to study these beautiful animals in the wild someday.

We lost track of time a bit while traveling, and as we approached the border with Cameroon, I realized that we had only half a month left in Africa. I couldn't wait to get back to Nastia, but I knew that I'd miss Sarit and Alex. By now I understood how incredibly lucky I was when they answered my online call for volunteers a year earlier. I could have searched for a hundred years and never found such perfect companions.

Sarit was a particularly unique character. In all that time she never complained about the difficulties of the journey. Somehow she managed to remain upbeat and outgoing even during dust storms. And she could be extremely persuasive. Given enough time, she would probably be able to talk a death camp commander into setting all the inmates free. Alex and I weren't particularly talkative and never found it easy to approach people we didn't know. Having Sarit with us made our life in Africa so much easier.

As for Alex, he alone replaced a whole crew of experts. If he ever had a business card, it would probably say something like "Alex: I solve problems." A quiet introvert, he seemed to be the opposite of Sarit. He could spend hours never saying a word, just walking around fixing things, from computers to clothing.

"Hey, guys," I said as we were waiting in line to get our passports stamped by Cameroon border guards, "remember how a year ago we talked about maybe going to Indonesia after Africa?"

"Yes."

"Are you sure you still want to go? We could meet sometime in June and travel together until September."

"Sure, why not?"

How wonderful, I thought. I'll still have my dream team for the next big expedition. Nastia will join us this time. Could it get any better?

30
Shades and Shadows
Osteolaemus tetraspis

Every exit is an entry somewhere else.
—Tom Stoppard

Cameroon is sometimes called "the miniature Africa" for its exceptional diversity. It has rain forests and mangroves in the south, seasonally flooded savannas in the center, and dry grasslands in the far north, as well as mountains and a large active volcano. It is one of very few countries in West Africa where wildlife still survives in good numbers. Unlike most national parks in Central Africa, those in Cameroon can be visited without paying hundreds or thousands of dollars for airplane charters and luxury lodges. No wonder Cameroon has always been popular with naturalists. I'd wanted to visit it since reading Gerald Durrell's books about this country as a kid. But this time it wasn't going to be the journey of my dreams, for exploring the country well would take months, and we had only two weeks left. And we had to get some data on dwarf crocodiles, because otherwise our entire trip to both Congos, Gabon, and Cameroon would be a very expensive disaster.

Recording the "songs" of dwarf crocodiles was critical for me. They are among the few crocodilian species adapted to living almost exclusively in small bodies of water. My theory predicted that they roar a lot, but seldom, if ever, headslap. Anecdotal information seemed to confirm this, but I needed firsthand data. Besides, just before our trip to Africa genetic studies had shown that "dwarfies" were not one, but three species. I'd hate to have such a huge hole in my data tables. One of these three species occurs only in northeastern Congo-Zaire; I saw it during my airplane trip to the Ituri Forest, but that was before I started my research on crocodilian "songs," so I didn't try to observe its behavior. Another species lives in far West Africa, from Senegal to Benin, and the third one inhabits Nigeria, Cameroon, and much of Central Africa. All three seem to be very similar in behavior and habitat preferences, so if I could get data on one, that would be good enough.

In addition to its exceptional natural diversity, Cameroon has over two hundred ethnic groups. It's the only country where Pygmies still speak their own languages. About a third of the people are Christians, a quarter are Muslims, but the rest still practice traditional religions.

This abundance of "pagan" souls to convert has made Cameroon a major battleground between Catholics and Protestants. Both communions have huge missions in major cities. We visited these missions a few times because they had the country's cheapest hotels. They reminded me of English castles built to dominate the unruly Welsh. Pope John Paul II had visited the country twice, and local TV programs include hours of hysterical sermons in addition to the usual mix of French pop music, Nigerian soap operas, and South African reality shows typical for sub-Saharan Africa. The middle class has largely been converted already; the rest of the population can't afford the tithes and continue worshipping gods of the forest, the sea, and Mount Camaroon. One popular belief is that you can stop a volcanic eruption by sacrificing five goats and one albino man. Albinos are unusually common in local villages, but every time there's an eruption of Mount Cameroon (they happen every few years) all of them flee to Nigeria to wait it out.

Cameroon's history has been relatively peaceful, but its cities often look like a civil war has just ended. The country is a typical fake democracy, with elections invariably rigged and the same people always remaining in power. As in almost all such cases, the quality of living is slowly deteriorating despite foreign aid and investment.

The English-speaking western part is essentially an occupied territory, annexed after a fake referendum. Despite being neglected by the Francophone government, this area is remarkably better organized and overall more comfortable to be in. In the rest of the country, expensive Greek bakeries are the main outposts of civilization among the sea of corruption and chaos.

Everything in Cameroon is a challenge, even taking a bus or taxi ride. Before getting in a car, you have to agree on the price *in writing*, make sure you have exact change, and let the driver know that you don't have a cent more. That still doesn't prevent you from being told halfway that the agreed price was only for going part of the distance. Once you've arrived at your destination and paid, expect the driver to come back two minutes later and claim that you haven't. He can gather a substantial crowd to back up his demands. But that's not all: expect to be hit for bribes at every police checkpoint (even if you are a bus passenger) and to get into a fatal crash sooner or later. Local drivers are so reckless that roads are lined with signs saying things like FIVE PEOPLE DIED HERE, or, in one case, SIXTY-FOUR PEOPLE DIED HERE.

On the bright side, you can expect all kinds of perks you don't get on Greyhound buses. In one bus we watched a traveling dentist walk down the aisle with thongs and a bottle of anesthetic, offering everyone to pull teeth. The road was bumpy, so when someone agreed to become a patient, three other passengers had to hold the dentist in place while he was doing his delicate job.

We crossed Cameroon from east to west and checked three nature reserves, but they didn't have crocodiles. One had a wildlife rehabilitation center where we met an orphaned baby chimp with a beautiful, almost angelic face. Ape faces are as easy to recognize as ours; two years later, I saw a poster with a photo of a baby chimp and instantly recognized that face.

The last reserve we had time to visit was Korup National Park on the border with Nigeria, one of the largest, best-protected, and most scenic forest reserves on the continent. African rain forests usually have lower diversity and smaller trees than similar forests of Asia and South America, but Korup is an exception. Some trees are almost two hundred feet tall. The forest is very dense and mysterious-looking, with quiet blackwater streams, strange flowers, and "groves" of bizarre mushroom-shaped termitaria. We saw few large animals, but birds were everywhere, and at night there was a frog or some other creature on every tree branch.

I spent nights watching small forest ponds. Our guide, a former poacher with impressive knowledge of local wildlife, told me that a sure way to know if a pond had crocs was to check for the presence of a small black-and-red bird called a blue-billed malimbe. I remembered that the only time I saw this bird in Gabon was at the only pond where I found a crocodile. I checked this method in Korup, and it worked! Later I learned that zoologists had already described that connection. They thought that the bird preferred to nest in branches hanging over the ponds with crocodiles because the crocs would repel or eat snakes and other predators. In one case a pair of malimbes nested near a concrete pool with dwarf crocodiles in a city zoo. I observed birds near crocodile ponds at the time when they weren't nesting, indicating that they might get some other benefits from the crocs' presence as well. For example, there could be more insects around such ponds because crocodiles eat fish so the fish eat fewer insect larvae.

The crocodiles I found were charming creatures, so dark-colored that they seemed to be made of black plastic. One was still a juvenile, but the other was a full-grown male, five feet long. Both were extremely shy, completely nocturnal, and remained hidden under tree roots or piles of fallen branches most of the time. When they did show up on the surface of the ponds, I had to watch them without moving a muscle; even the slightest movement would make them disappear.

My patience was rewarded only once. The male raised his head and tail and gave a low-pitched growl that sounded like something a drunkard would utter before collapsing under the table. I was sure I saw his back vibrate in the split second before the sound, indicating that infrasound was also produced. Then he looked up, as if making sure his "song" didn't draw unwanted attention. I could only wonder to which extent the behavior of this crocodile had been modified by the constant threat of being shot.

Even in Korup there was a lot of poaching going on, mostly with wire snares, but also with guns. No park in Africa is completely free of poaching. After Korup we went to Mount Cameroon for a day and a half to have a break from the hot lowlands, and, as we were climbing to the tree line, met an armed guy carrying down a dead bushbuck (a small antelope). He was totally relaxed when he saw us. We talked to him and found that his main job was working as a tour guide for a company named something like Friends of the Forest Sustainable Ecotours.

A few years later, I was giving a talk about crocodilian behavior at the Cornell Lab of Ornithology in Ithaca, New York. After the talk, Cornell researchers asked me to help them identify the mysterious sounds they had recorded with automatic microphones at *bai* (forest clearings where animals come to dig for mineral salts) in northeastern Gabon. They were studying forest elephants, so they used microphones sensitive to low frequencies. The moment I looked at the spectrograms I knew these sounds had been made by dwarf crocodiles. After so much time wasted in Central Africa to get just one observation, I suddenly had dozens of good-quality recordings, clearly showing bellows and infrasound, but no headslaps. The biggest gap in my data was closed just when I thought it would never be.

On our last day in Cameroon we went to see rare African manatees in a small lake near the coast. The entire lake was crisscrossed with fishing nets, but somehow a few manatees survived there, as did a small crocodile. I took it for a Nile croc and was glad to see it in Cameroon, where its species had almost completely been hunted out. Only later, when the paper about the discovery of the sacred crocodile was published, did I realize that the croc I saw was the only sacred crocodile I had ever seen in the wild. I should've paid more attention to it.

Sarit and Alex decided to go to Morocco to get more data for our guidebook. I admired their perseverance: we all felt badly overdosed on Africa. As for me, I needed to get home. The alligator mating season had already begun, and I was also missing Nastia terribly. We had known each other for a year and a half, but of that time we'd only been together for about a month.

We were waiting for our flights at the airport when suddenly we heard Russian speech. A dozen men were standing near a pile of luggage, looking lost. We asked if they needed help. They were Latvian seamen who were supposed to start working on a cargo ship in Cape Town. They booked tickets through a Russian company, arrived in Cameroon to change flights, and discovered that the company had pocketed the money for the second leg of their route. We helped them leave the airport without getting ripped off in five different ways, but we didn't know if they were able to complete their taxi ride safely. In the process they gave us some particularly virulent strain of flu that made us seriously sick for a week and later jumped from me to Nastia, spoiling our next short stay together.

Sarit and Alex left for Morocco, while I still had to go through a few hoops to get myself and my new viral companions back home.

If there's one thing I miss about the Soviet Union, it's the state-owned airline, Aeroflot. It had fixed prices that remained unchanged for years. All tickets were refundable. I still remember that the largest fare, for the seven-hour flight to Chukotka, was 198 rubles (about $120), which wasn't bad. As a schoolkid I used to fly to Turkmenistan for spring and fall breaks to catch snakes for snake farms and could make enough money in two weeks to travel to the Russian Far East in the summer. These flights were always fun, you could get rare views of the remotest parts of Siberia, and the dinner often included smoked salmon, caviar, and tundra blueberries.

Nowadays air travel is anything but fun. You have to spend a whole day online to find the cheapest rate, rates vary day to day, prices aren't proportional to distance, all but the most expensive tickets are nonrefundable, and direct flights often aren't the cheapest option. I found a way to get from Cameroon to Washington for half the price of a direct flight, but it involved five days of travel, six plane changes, and covering three times the distance.

I flew to Libreville and then to Ethiopia, where I had to wait for three days. I managed to rent a small car that I wasn't supposed to drive out of town. But Addis doesn't have an official city limit, so I could stretch it a bit and explore the forests of giant juniper to the west. It saved me money, because I could sleep in the car instead of paying for a hotel, and food was much cheaper in the countryside.

In the airport I met a guy I knew, a zoologist studying gelada baboons. We were supposed to be on the same flight, but he got thrown out for trying to take a bag of chili peppers on board. He didn't know that a special rule prohibited transporting peppers. The original regulation was probably about pepper spray, but one word got lost in translation.

The next plane change was in Khartoum, the capital of Sudan. The city didn't look like much, but it was nice to see the confluence of the White and Blue Niles from the air. A small plane took me to the United Arab Emirates, where I enjoyed an aerial view of Burj Khalifa, the world's tallest (over half a mile high) and most impossible-looking building. Then another landing in Amman, a brief view of snow-clad mountains as we flew over Lebanon, a plane change in London, renting a car in Washington, and a long, long drive to Tennessee.

It was early May. The farther south I drove the more green grass and young leaves I saw. Even the pavement, wet from light rains, was trying to be colorful by reflecting blue sky and an occasional rainbow. I was smiling the whole way. I knew that beyond the mountains a girl with green eyes was waiting for me, beautiful and tender like the Appalachian spring.

31
Moving Home
Alligator mississippiensis

Facts are the air of scientists. Without them you can never fly.
—Linus Pauling

YET AGAIN, I COULD SPEND ONLY A FEW DAYS WITH NASTIA. Leaving her was getting increasingly difficult, but I had to get to Florida as soon as possible or I'd miss the alligator mating season.

I decided to do the same thing I did with Nile crocodiles at Lake Turkana: compare "songs" of gators living in a big lake and in small ponds nearby. The study site I chose was Ocala National Forest on the shores of Lake George.

Ocala National Forest is mostly dry pineland on sandy soil. Small lakes are either sinkholes or "gator holes" dug out by many generations of alligators. In some places there's a layer of clay under the sand, so after rains the dunes become saturated with water and turn into quicksand. My car got stuck so badly once it took me six hours to free it by jacking up its wheels one by one and sticking tree branches underneath.

The results of my Ocala study were very clear. Just as in the case of Nile crocodiles, alligators living within the same area had exactly the same "songs" no matter whether they lived in Lake George or in small ponds nearby. This finally confirmed that individual crocodilians can't change their "songs" in response to their habitat. Differences in "songs" between populations that I discovered are genetic and probably take a long time to evolve.

It was already very hot, and horseflies were emerging in frightening numbers when I was finishing my work in Ocala. Fortunately, this time I didn't have to sleep in my car. I had a couple of good friends living nearby. My books, furniture, and other stuff were kept in their garage. Every time I was about to drive from Florida to Tennessee, I loaded my car with whatever I could squeeze in, gradually transporting my things to Nastia's place.

Nastia was surprised to see what I was bringing in. A traditional subject of Russian women's complaints about men is the latter's alleged tendency to leave dirty socks scattered around the apartment. Nastia soon realized that it would never be a problem with me. I had very few clothes left after so much time in the field, and I seldom use socks because I wear sandals unless it's freezing. Instead, I brought to her place a lot of books, pole nets, mist nets, Sherman traps, binoculars, a kayak with oars, a wetsuit, a diving belt, two sets of fins, a telescope, an old Zeiss microscope that my family had managed to save during two world wars, some beautiful seashells I had collected a long time ago, before shell collecting became a serious environmental problem, and lots of small things, from shark teeth to UV flashlights for scorpion hunting.

Meanwhile Nastia was busy finishing her semester and taking a scuba diving course. We knew that she would have to dive a lot where we were going that summer. I returned to Knoxville just in time for her final diving lessons, taken in an old rock quarry. On that day there was a supercell thunderstorm, so in addition to very cold water Nastia had to endure driving rain, large hail, and gale-force wind before getting underwater. But she passed the test with flying colors and enjoyed the experience. "Little Mermaid," the nickname I'd given her when we first met, suited her very well.

She was now free to join me in Florida and we were happy to be on the road together.

In Miami I met with my advisor Steve to report my results. Again, Steve thought that I was making good progress, while I wasn't sure about it at all. I had ever-growing piles of data on two-thirds of the world's crocodilian species, but very few answers.

Nastia and I drove north along the western coast of Florida. We stopped at Crystal River to snorkel with manatees and continued to Louisiana. I had to finish gathering data from my study site in the Mississippi Valley, where I hadn't completed it the previous year.

By that time the summer rains had arrived. We barely missed a tornado on a causeway to New Orleans. I love storm chasing and would have enjoyed following that storm front for a few days, but we had to hurry to catch the last week of alligator "songs."

One stormy night we were driving through a remote wildlife refuge and saw that the road was covered with frogs. I took a few inside the car to photograph. We were busy taking pictures when a police car stopped behind us. I've noticed that cops in the States never miss a chance to look inside a car parked in the woods, even if there's no reason to suspect any illegal activity. "What are you doing?" the officer asked, as if expecting us to admit that we were drawing plans to blow up Fort Knox, or, worse, having sex on public property. "Photographing frogs," we answered. The look of disappointment on his face reminded me of some of the policemen I'd met in Mozambique.

Encounters with suspicious police officers are a regular experience for many field biologists. Once I was pulled over when I was driving late at night on an empty road with lots of migrating newts. The police apparently found it strange that I was driving in a zigzag. I tried to explain that I was trying to avoid squashing newts, but they made me take a barrage of tests. After enduring this for twenty minutes I finally got them to listen, and told them about the amazing life cycle of newts and the importance of their conservation. When the police car left, I was glad to see that it was zigzagging.

But not all such encounters end well. Just like me, Steve had spent some of his teenage years catching snakes for snake farms. Once he was returning home after a night of catching rattlesnakes and stopped at a gas station to have a cup of coffee. As he was sitting in his car sipping coffee, a policeman showed up and demanded to know what was in all those wooden boxes in the trunk. "Rattlesnakes," said Steve. "Stop bullshitting me! Open these boxes!" shouted the officer.

"I can't," said Steve, "there are snakes inside." The policeman drew his gun and, despite Steve's warnings, opened one of the boxes. A large rattlesnake raised its head. The officer jumped back and accidentally shot himself in the foot. Steve was prosecuted for endangering the life of a police officer. He still has his mug shot stapled to a wall in his office.

It took us only a few days to get the missing data in the flooded Mississippi Valley. This was my last research trip to the Deep South, an enchanted land I love. It takes time to figure out its subtle laws, to learn the ways of its inhabitants, human and nonhuman, and to discover its unusual beauty. The most important thing to understand is that in such a flat country, even the slightest differences in elevation can create different worlds. The dry pine woods with their slender trees; the shady oak forests filled with the aroma of magnolias; and the most unique and majestic of them, the cypress swamps, where everything is aquatic, even rabbits.

The alligators' mating season was over, and I had two weeks left before returning to the tropics. We drove to Lake Superior and circled Isle Royale on a yacht with our friends. It was so nice to travel just for fun, and not think about crocodiles all the time.

As soon as we returned to Knoxville, I had to leave Nastia again. She would join me a month later. This time my flight back to the tropics was from New York. To get there as cheaply as possible, I chose a rather unconventional way. I used my connections in the Hidden Empire.

32
Ghost Hunt
Tomistoma schlegelii

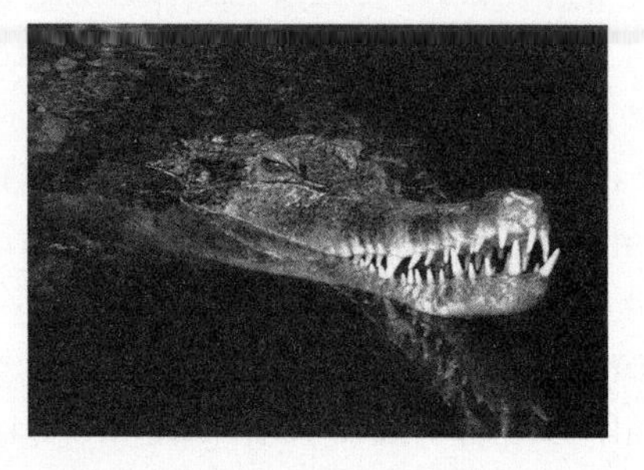

Each problem that I solved became a tool that served afterwards to solve other problems.
—René Descartes

FOR MOST AMERICANS, THEIR OWN COUNTRY IS AN EXPENSIVE ONE. They pay at least $40 (usually a lot more) for a night in a hotel, shun Greyhound buses because long-distance bus rides cost nearly as much as low-cost flights, and honestly believe that cheap food means junk food.

But Chinese Americans live in better style. They go to Chinese restaurants where waiters don't speak English but the food is almost as good as in Sichuan and costs almost as little. They use cheap Chinese hotels with excellent service. And they have their own long-distance bus network.

Nastia had a Chinese friend named Peng who trusted her enough to show her a bit of this secret world. She would take Nastia to restaurants where Peng had to argue with waitresses every time because they didn't want to serve Nastia the best dishes, claiming that a round-eyed barbarian wouldn't be able to appreciate them. She also told us about the clandestine bus network. To get a bus ticket, you had to call an operator who spoke only Mandarin, and show up in some remote, poorly lit parking lot to be picked up. You also had to know a few

Chinese words yourself to talk to the driver. But once you were on the bus, you found fully reclining chairs, a choice of food from different provinces, and good satellite Internet. You got from Knoxville to New York much faster than by Greyhound and for less than half the price.

I met Sarit and Alex in New York, and we flew to Jakarta via Anchorage, spent a night in the airport, and continued to Medan in North Sumatra. Our plan was to study two of the four crocodile species occurring in Indonesia.

In the Americas there's one widespread coastal species of crocodile, the American crocodile, and three freshwater croc species with much smaller ranges. In Asia and Australia the situation is remarkably similar: there's one widespread coastal species, called the saltwater crocodile, and five freshwater crocodiles with more localized ranges. By that time I had studied only one of these species, the Indian mugger, so there was a lot of work to do.

But there was yet another Asian crocodilian that I wanted to find and observe even more. It was a strange creature called the false gharial. Recently, many zoologists have switched to calling it *Tomistoma*, the scientific name of its genus, of which it's the only living member. They want to avoid the implications contained in "false gharial," because nobody knows what it is: a gharial that looks like a crocodile, or a crocodile that looks like a gharial.

For centuries, classifying plants and animals was based on their morphology (shape and structure). Since the time of Darwin, it was assumed that such a classification would reflect the evolutionary history. Two species classified as being close to each other would have a recent common ancestor. The higher classification levels—genus, family, order, class, phylum—would correspond to larger and larger branches of the tree of life.

This approach has inherent problems. You don't always know which features are more important than others. Crocodiles, for example, look a bit like big lizards, but once you look at their anatomy, physiology, and embryonic development, you realize that they are closer to birds. Sometimes two kinds of organisms that aren't closely related can look almost exactly the same because their lifestyle is

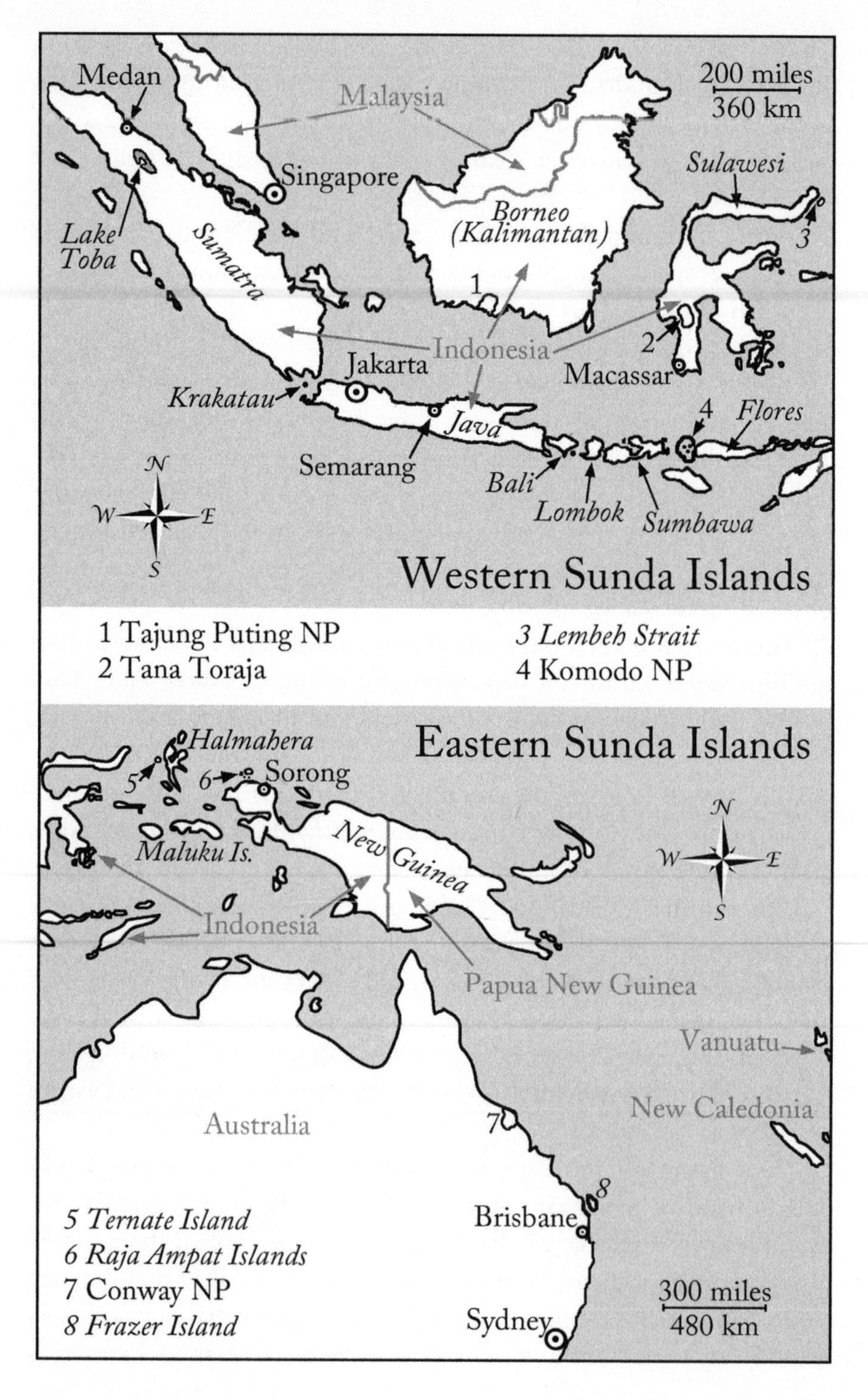

Medan
Malaysia
200 miles
360 km
Singapore
Sulawesi
Lake Toba
Sumatra
Borneo (Kalimantan)
3
1
2
Indonesia
Jakarta
Macassar
Krakatau
Java
4
Flores
Semarang
Bali
Lombok
Sumbawa
N
W
E
S
Western Sunda Islands
1 Tajung Puting NP
2 Tana Toraja
3 Lembeh Strait
4 Komodo NP
Eastern Sunda Islands
Halmahera
5
6
Sorong
Maluku Is.
New Guinea
Indonesia
Papua New Guinea
Vanuatu
New Caledonia
Australia
7
8
Brisbane
Sydney
300 miles
480 km
5 Ternate Island
6 Raja Ampat Islands
7 Conway NP
8 Frazer Island

similar. Moles of the Northern Hemisphere, golden moles of Africa, and marsupial moles of Australia belong to three unrelated lineages of mammals, but a nonexpert would have a hard time telling them apart. Biologists call such similarity between unrelated organisms *convergence*, meaning that two twigs from distant branches of the tree of life converge on the same design.

The false gharial is a large fish-eating species living in rivers. It looks a lot like the Indian gharial, with a powerful tail and long, narrow snout. But its teeth, skull structure, leg anatomy, and numerous other features have a lot more in common with crocodiles than with gharials. So it was assumed that its superficial similarity to the Indian gharial was due to convergence and that the false gharial was, in fact, an aberrant crocodile.

But in the last decades of the twentieth century a revolution in the science of biological classification took place. New methods allowed for comparing the molecules making up living organisms, and then, as the new techniques became even more sophisticated, their genetic codes. At first it seemed that the new approaches would give us the absolute truth, exposing the real evolutionary relationships between organisms. But many of the first molecular results were later proven wrong. There's still no way to find an "absolute truth" about the relationships between many groups of animals and plants. The overall shape of the tree of life is gradually becoming much clearer than just twenty years ago, but there are many difficult cases where different methods produce conflicting results.

The false gharial was one such case. All molecular studies showed it to be a close relative of the Indian gharial rather than crocodiles. But that was the opposite of what morphology suggested. Paleontological data wasn't much help: it was known that false gharials had marine ancestors, but the earlier stages of their evolution were obscure.

It occurred to me that I could provide a decisive argument in this controversy by studying the behavior of the false gharial, which would be a fourth source of information, independent from morphology, fossils, and molecular data. I knew that postures and "songs" of the Indian gharial are strikingly different from those of crocodiles. The gharial has the snout-up posture, "pong" sounds, and "buzzes."

Crocodiles have a posture in which they raise both their head and tail, and their "songs" include infrasound, roars, and headslaps.

All I had to do was observe the behavior of false gharials during the mating season. But they are rare in captivity and on the verge of extinction in the wild. Once common in Indochina and the western part of Indonesia, they now occur only in a handful of places in Malaysia, Sumatra, and Borneo. We started in Sumatra.

Indonesia is a very cheap country to travel in, as long as you stay away from overpriced tourist attractions and from the far eastern part, where infrastructure is poor and you have to fly to get anywhere. People are exceptionally nice: quiet, friendly, and polite. The country has outstandingly high biodiversity, which is now being rapidly destroyed by deforestation, overfishing, and the illegal wildlife trade. What remains can still be seen in national parks, but most parks have recently been stricken by the "obligatory guide" plague, which has emerged from Africa and spread over all of India and much of South America.

Just a few years ago you could come to any national park in Indonesia, pay a reasonable entry fee, pitch your tent in the forest, and walk around looking for animals. Nowadays, you must be accompanied by a guide, who often speaks little English and doesn't know much about plants and animals but charges more in a day than a local farmer makes in a week. There are exceptions, but even the best guide is not welcome company because you are forced to hire him. This extortion scheme is supposed to provide income for locals, compensating them for not logging the park and not killing everything there that moves as they'd normally do. But this never works. A few people such as hard-core bird-watchers are always ready to pay *any* money for a chance to see a rare bird, but there's never enough of them to buy out even one acre of high-quality tropical timber. Other tourists just move to places where they aren't subjected to a racket. No wonder tourism in Indonesian parks is declining, while in the parks of neighboring Malaysia, where guides aren't obligatory, it's flourishing.

For us this obligatory guide scheme was a disaster. We couldn't afford those insane expenses. Besides, false gharials are known to be extremely shy in the wild, and the chances of finding, not to mention observing, these ghosts of the river while being accompanied by a noisy guide would be zero. We tried four national parks on Sumatra. In three of them the habitat wasn't right; in one park false gharials were still present, but the combined entry, camping, guide, and boat fees would consume our summer budget in three days.

But those three weeks on Sumatra weren't a total loss. We got to see Lake Toba, a ring-shaped volcanic lake more than fifty miles across, created seventy thousand years ago by an eruption so cataclysmic that it probably caused something like an "atomic winter" and almost killed off the entire human race. And we became the first naturalists to see a striped rabbit in the wild.

Striped rabbits are rare animals of montane rain forests. One of them is known from a few specimens trapped in Sumatra; the other has been recently discovered in photos made by automatic cameras in Laos and Vietnam. We ran into our rabbit accidentally. Late that night we were riding in a car we shared with some fellow travelers. I remember thinking that the habitat seemed right for striped rabbits but the chances of seeing one were zero. And then we saw one sitting on the roadside. Everybody shouted, "Striped rabbit!" at the same time. The driver, who thought we were all asleep, was so startled that he almost lost control of the car. Normally, I wouldn't even try publishing a sight observation like this without solid evidence, but this time I was lucky to have five people in the car with me and could list them as witnesses.

We gave up looking for crocodiles on Sumatra and flagged down a southbound truck loaded with oil palm nuts. The driver agreed to let us sit in the open truck bed for the entire ride to the southern tip of the island. It took thirty hours, with a freezing night on a mountain pass, two hot days in the sun, and a few rain showers. But it was a pleasant ride except for the worms. The palm nuts were crawling with small beetle larvae. Even two weeks later we were still finding those brown grubs in our clothes, hair, and backpacks.

Before crossing to Java, the next big island, we made a short side trip to Krakatau, the famous volcano that exploded in the late nineteenth century, killing about a hundred thousand people and causing a "year without a summer" worldwide. The crescent-shaped remnants of the old cone form a broken ring around the new island, called Anak ("child") Krakatau, which has grown in the center of the explosive crater during the last hundred years. Anak Krakatau produces small eruptions every few minutes or hours.

It was difficult to find boatmen willing to spend the night on Anak, and when we got there we understood why. It's an eerie place. The island is surrounded by still, dark waters of the deep bay enclosed by the black vertical walls of the old crater fragments. The beaches are also black and the thin band of jungle is covered with the black ash that rains from the sky after every explosion. Even ghost crabs on the beach are black. The central part of the island is a dead zone, regularly bombarded with red-hot rocks up to the size of a refrigerator.

We climbed to the rim of the new crater and spent the night watching the explosions up close. Every time the entire island trembled briefly, and a fountain of molten rocks shot into the night sky. The rocks flew in beautiful parabolas and fell all around us, breaking into shrapnel and kicking up clouds of ash. To our amazement, we noticed that the crater rim wasn't completely devoid of life; there were a few small trees with broken branches and torn-off bark, some earwigs under rocks, and a pair of nightjars hunting for moths flying toward the lava.

The spectacle was certainly worth the risk. If there's a sight on our planet worth seriously risking your life for, it's a volcanic eruption.

I went to Jakarta to meet Nastia at the airport. This was her first long trip to the tropics. I wanted to give her a chance to acclimatize at higher elevation, where it was cooler, so I took everybody to a national park in the mountains. The park was beautiful, but the rest of the plan didn't work well. Heat wasn't a problem; Indonesia has a wonderful climate, and it was never too hot for comfort during our three-month stay there. But the food was difficult for Nastia to get used to. She got sick almost immediately and took almost three weeks to recover. She tried to remain upbeat, even when she was so weak that she could barely walk. She became very thin, and looked even more like the Little Mermaid.

Our next place to look for false gharials was Borneo, called Kalimantan in Indonesia. The northern part of the island belongs to Malaysia and has some of the best national parks in the world, but it's mountainous, so large jungle rivers are few and false gharials are vanishingly rare. We decided to go to the Indonesian part.

First we had to get from West to Central Java. The islands of Indonesia look small on world maps, but because of all the mountains, cities, and rivers, traveling overland is excruciatingly slow, even on Java, where highways aren't that bad. It took three days to reach Semarang, the city from which the next ferry to Kalimantan was due to sail. I was glad we had to go through Semarang because I wanted to see one of its temples.

Semarang has the largest Chinese community in Indonesia. The Chinese diaspora fills the same niche in Southeast Asia that Jews used to fill in Europe. They are well-educated, entrepreneurial people who conduct much international commerce, travel widely, and trade in everything. The similarity with Jews is remarkable: there are even occasional pogroms in Indonesia and elsewhere. Being the nation of travelers, the Chinese of the diaspora have a well-fitting deity.

In 1405, the Chinese emperor sent a huge fleet to explore the world, commanded by Admiral Zheng He, a brilliant navigator and geographer. The fleet made seven voyages, visiting Africa, Arabia, Persia, Sri Lanka, India, and probably New Guinea and Australia.

All those explorations were in vain. The next emperor banned overseas expeditions and closed the borders. But Zheng wasn't forgotten. He is worshipped in many parts of Asia as a saint under the name Sam Poo Kong. Gedung Batu, the largest temple built to honor him, is in Semarang. It looks like any modern Chinese temple, with kitschy stone dragons, gypsum lions, and paper lanterns. A tunnel leads to a grotto with statues of Zheng and his two captains decorated with flashing electric lights.

But there's something special about this temple. Maybe it's Zheng's face on his numerous portraits, unusually young for a Chinese saint.

Maybe it's the sails; there are hundreds of junks pictured on the walls. Or maybe it's the fresh wind, which reaches the temple grounds from the beaches on the other side of the city despite all the smog: the wind of faraway seas and lands.

We wanted to go to a national park called Tanjung Puting. Ferries are the main mode of transportation between Indonesian islands. We got tickets for the cheapest deck and set up tents there. The crossing to Kalimantan took two days, during which Nastia was the center of attention. Many people had never seen a blonde before, so they constantly photographed her, even when she was sleeping in the tent, and then proudly showed her the pictures. Women were trying to rub elbows with her, believing that their own skin would become lighter from contact. Light skin is considered beautiful everywhere in the tropics, probably because limited exposure to sun is associated with higher social status.

It took another day to get park permits, register with the police (which required a trip to another town), and find a boat captain willing to pretend he was also a guide.

The river journey was wonderful. At first we were moving up a whitewater river (with milk-colored water) lined with mangrove palms. Farther upstream the "normal" mangroves began. They were inhabited by hundreds of proboscis monkeys and crab-eating macaques; they'd often jump from tree to tree across the river, or drop in the water and swim frantically to the other side. At the end of the day we turned into a small blackwater stream and followed it until it became too shallow. Now we were in *kerangas*, or peat-swamp forest. It's a kind of jungle growing on acidic soil, with small trees, numerous grassy openings, and lots of pitcher plants. There are few large animals in the kerangas except for leaf monkeys, gibbons, and orangutans. There was an orangutan rehabilitation center nearby, so many of the apes in the forest were absolutely tame. Mother orangutans seemed to like Nastia more than other people, probably because she was very quiet and gentle. They'd even offer her to play with their babies, but we had to be careful because baby orangutans can easily contract human diseases.

That blackwater river was known to have false gharials. A year earlier a tourist had been killed by one, which was unusual, because false gharials are mostly fish-eaters. But I didn't see any of them. Our boat was too big and noisy. I wished I had my kayak with me, but transporting it around Indonesia would've been a nightmare. Desperate, I went to the research station manager and asked to borrow a small canoe. To my great surprise, he agreed.

I spent countless hours paddling quietly up and down the black forest stream. There were beautiful birds and butterflies during the day, but nights were even better. Fish owls watched me from trees. Eyes of giant, fully transparent ghost shrimp reflected the flashlight from underwater like hot coals. Sometimes the flashlight beam would fall on a sleeping kingfisher, a shining turquoise-and-red gem the size of a plum.

Those false gharials were the most cryptic crocodilians I've ever had to look for. I couldn't find them even when floating downstream as quietly as I could, using the paddle only for steering. Once I spotted a cute hatchling about ten inches long, handsomely spotted, with chopstick-like jaws. But even that tiny thing disappeared underwater before I could get a good photo.

There was a shallow pond connected to the river in one place. I knew that many crocodilians prefer shallow water for "singing," so one evening I left the canoe far upstream, swam into the pool, and waited, sitting neck-deep in the water. After midnight a huge false gharial, almost twenty feet long, suddenly surfaced in front of me. I don't know how he managed to get there without moving a single grass blade. He remained motionless in the moonlight for a full hour. I was so afraid he would notice me and vanish that I hardly dared blink. Then I noticed that he was floating high in the water. Crocodilians can do that by inhaling air; such a "swimming inflated" posture sometimes leads to a "song" but not always. A few more minutes, and the giant raised its head and tail tip—a sure sign that he was about to "sing." I watched and listened so intensely that I didn't blink at all; my eyes began to hurt. I could hear every mosquito around me and hoped that not too many would bite. Finally, the "song" came, but it was only infrasound. The vibration was so intense that I felt like a victim of dynamite fishing. A tree growing above my head showered me with dry leaves. The giant remained still for another hour, then his mighty tail slowly moved, and he slid through the grass toward the river and disappeared.

The next night I was there again but didn't see anything. And then we had to leave. Renting the big boat was too expensive.

A year later I got firsthand observations and eyewitness accounts of the "songs" of false gharials kept in zoos. In addition to the infrasound, males often headslap but never roar except when they feel threatened. This matches the predictions of my theory: since these animals inhabit only rivers, they don't need roars to tell others where they are. Infrasound and headslaps can carry through the water much farther than roars can through the air.

The "songs" and postures of false gharials are typical for crocodiles and totally unlike those of Indian gharials. So the "classic" systematics is probably right, and these rain forest giants are closer to crocodiles.

33
Island-Hopping
Crocodylus porosus

When you come to a fork in the road, take it.
—Yogi Berra

THE NEXT SPECIES TO STUDY, AND THE ONE I was particularly looking forward to meeting in person, was the saltwater crocodile, or "saltie" as it's sometimes called in Australia—the largest living reptile. Exceptionally large males can be over twenty feet long and weigh more than a ton. Behaviorally, it's the closest thing we have today to the marine crocodiles of the Jurassic. Saltwater crocs have been seen in the open ocean a thousand miles from land. They have colonized remote oceanic islands, such as Palau, and shown up in Japan. One small population in Australia was recently discovered to be migratory, spending summers at Fraser Island and swimming two hundred miles north to escape the winter cold. These are extreme cases, but most salties cover a lot of ground (or, rather, a lot of water) in their search of permanent homes during adolescence.

Not surprisingly for such a great traveler, this species was once the most widespread of any crocodilian, occurring from the eastern coast of Africa to Fiji. But one of its adaptations to long-distance travel is its relatively light armor, with few bony platelets embedded in its belly scales; this makes its skin highly valuable, so it has been hunted out in many parts of its former range. It's extinct in Africa and Fiji; there are only a few populations left in mainland Asia and only three individuals in Vanuatu. Most remaining salties, probably about two hundred thousand, live in northern Australia.

Up-to-date information on croc populations was difficult to obtain, and published maps still showed continuous distribution from India to Australia, so I expected salties to be easy to find in Indonesia. But we couldn't find them on Java, and I later learned that there was only one tiny population left there. We tried Bali, Lombok, Sumbawa, and Flores, and everywhere local people informed us that it had been years since the last croc sightings. In some places croc safety rules had morphed into taboos that were no longer reasonable, like in one village on Lombok where people still believed that anyone bathing in the river at night would be consumed by water demons with sharp teeth. We kept island-hopping, occasionally stopping to explore the forests and mangroves (salties' favorite habitats are mangrove lagoons and river estuaries) or to scuba dive if we got tired of traveling.

It wasn't a relaxed vacation. Even small islands took forever to cross overland, not to mention explore. Flores is two hundred miles long, and most of its highways are paved (unlike on some other islands), but covering half of its length by bus and hitchhiking took three days. A ride scheduled to take four hours could easily take eleven, because the driver would frequently stop to eat, wash himself in front of everybody, chat with friends, or load and unload endless boxes of stuff he was delivering.

The atmosphere in our team wasn't relaxing either. Before bringing Nastia on our first long trip, I was confident in two things. First, I knew I had perfect travel companions. Nothing less than a serious illness could distract them from our journey, adventures, and the beauty we were seeing. They never had unreasonable complaints, petty issues, or unjustified worries. Second, since Nastia was so charming and easygoing, they'd fall in love with her immediately, just as I did.

Both assumptions were wrong. For the first time in our year-long journey and friendship, Sarit was upset for no obvious reason. She didn't like Nastia being around, and she didn't like me paying so much attention to her. I don't think it had anything to do with

Nastia's personality. My friends just worried that Nastia could take me away from our nomadic lifestyle.

Sarit tried to persuade me that Nastia was a wrong girl for me. It wasn't a good idea, because even if I didn't love Nastia as much as I did, I'd never consider leaving her alone in the Indonesian boondocks. I hoped that things would get better with time, but after a few weeks Sarit insisted that we split one way or another.

It was a sad good-bye. My friends and I had been through so much together. We shared not just food and shelter but all our memories and dreams during our journeys. But I couldn't give up my life just to be able to keep traveling with them.

Sarit and Alex went to Malaysia. Nastia and I continued our journey east, from island to island, looking for crocodiles but never finding them. Another week went by. We were getting desperate as our precious summer was being wasted on endless buses, trucks, and ferries, with nights often spent in police offices at bus or ferry terminals. The police were always hospitable, fellow passengers always merry, and sometimes the entire bus would sing song after song for hours. Soon Nastia learned enough Indonesian to join the chorus.

Our only consolation was diving. The so-called Coral Triangle, the area from Bali and the Philippines to Fiji, is where all modern coral-reef fauna originates from, so the diversity there is mind-blowing. Diving in these waters is an experience completely different from the much younger reefs of the Caribbean or the Red Sea. An hour-long dive feels a bit like having half a day to explore all the museums of Rome. But you don't need to dive to notice that all other seas are dull and boring compared to these enchanted waters. From the deck of a boat, you can see the colors of fish fifty feet below the surface. Even the sea itself is more colorful than you've ever thought water could be—an intense, ever-changing mixture of turquoise and blue. Of course, since you are less than a thousand miles from Asia, you have to learn to ignore countless floating plastic bags and torn fishing nets. But at the present rate all of the world's seas and oceans will soon be covered with garbage, so one had better get used to it.

Unfortunately, many parts of the Coral Triangle are being destroyed by dynamite and cyanide fishing. The latter method poisons all living things around, but a few fish are stunned rather than killed, and can be collected for pet trade. Most of them later die in transit, but one out of a few hundred survives. That's where the fish you see in saltwater aquariums come from. Sharks are caught for shark fin soup, and sea horses for a potion popular in much of Asia as a divorce prophylactic.

We particularly enjoyed a boat trip to Komodo National Park, a great three-day break from our hectic run across Indonesia. The park is a world of small volcanic islands surrounded by calm sea, with coral gardens underwater. Once we saw a particularly impressive sunset there. I said something like "All that is missing is a green flash"—and the next moment the upper edge of the solar disk turned emerald-green just before disappearing below the horizon, turning the sky and the sea from purple to silver for a second or two.

But even in that paradise there were snakes. When we stopped at one of the islands to have a look at Komodo dragons, a park ranger offered to sell us a gorgeous baby python of a particularly rare species, protected by both Indonesian and international laws. He wanted $2,000 for it. A local photographer who was traveling with us on the boat was desperate. "It's such a shame I don't have a British visa right now," he lamented. "All I'd have to do is put this baby under my shirt and get on a plane to London. I know a wildlife dealer who'd pay me five times that much!"

Probably the most interesting island in Indonesia is Sulawesi. It has never been connected to Asia or Australia, so its wildlife is mostly unique and strange. The K-shaped island looks weird even on the map. We decided to fly there, but it wasn't simple. My tourist visa to Indonesia was about to expire, so I had to leave the country at least for a few hours (yet another typically wise and logical visa-related rule). We returned from Flores to Bali, Nastia flew directly to Sulawesi, and I had to go through Malaysia. Just two more days of life wasted. No big deal.

Like almost all Indonesian islands, Sulawesi has suffered terrible deforestation in the last half century. You can find if there's any forest left around a village by looking at caged birds. Indonesians are crazy about birds and keep something like fifty million of them in tiny cages. Some species have been brought to the verge of extinction by rich locals looking for rare pets. If you see that all houses in a village have only common bulbuls, doves, and canaries, you know that all surrounding lands have been converted to fields and plantations. But if there are dozens of colorful species in those cages, there must be some forest left nearby.

Crossing Sulawesi southwest to northeast took us a week. The center of the island is the densely populated Tana Toraja region, famous for its strangely optimistic burial customs, which are remarkably similar to those of Madagascar and remote parts of the Philippines. The deceased aren't buried. Their bodies are first placed under a boat-shaped communal house called *tongkonan* and remain there for up to a few weeks while their relatives sacrifice numerous water buffaloes and feast. Then the remains are stored either in coffins hanging from cliffs or in burial caves. All around them, and in the houses where these people used to live, their relatives install the *tau-tau*, wooden effigies of the dead, skillfully made and strikingly realistic. The tau-tau mostly portray old people, so local villages sometimes look like sunny retirement homes for wooden dolls. The effigies have attentive or absentminded faces and sparse gray hair. Nobody is scared of them. They are loved, taken care of, and asked for advice in important situations.

We couldn't find any crocs on Sulawesi. Later Brandon Sideleau, a smart Australian scientist, conducted an extensive survey of local press worldwide, looking for reports of croc attacks on humans. Such reports seldom make it to English-language media. This way he obtained proof of the continuing survival of many croc populations believed extinct by zoologists, including some on Sulawesi. Perhaps I would've found Sulawesi crocs eventually, but it was almost the end of Nastia's summer break, so instead of subjecting her to more long-distance bus rides I took her diving in an unusual place called Lembeh Strait.

Lembeh has its share of beautiful coral reefs, but it's more famous as the world's best site for so-called *muck diving*. In a few places in and around eastern Indonesia, strange ancient fauna exists on shallows with black volcanic sands. These mollusks, fish, and crustaceans are so weird that it's often difficult to figure out what you are looking at. Most of them match their dark environment, but some are insanely colorful. This strange world has only recently been discovered; some of its inhabitants have names like "wonderpus" or "flamboyant cuttlefish," given by recreational divers who first encountered them.

There was one creature I wanted to see more than anything in Indonesia: the mimic octopus. Its discovery in the late 1990s was the strongest hint so far that highly intelligent life-forms might exist among the millions of species on Earth that we know nothing about. (Other recent candidates include certain jumping spiders and mantis shrimp.)

The mimic octopus is rare and seldom seen, but we got lucky. We found it on our last dive together. It is about the size of a fist (if it tucks all its arms in), usually blue and white, although, like most octopuses, it can change color better than any chameleon. What is so unique about it is that it can mimic at least fifteen different marine organisms, depending on the kind of threat or prey it encounters. It can instantly turn itself into a sea snake, a flounder, a goby, a mass of floating seaweed, a lionfish, a stingray, a brittle star, a jellyfish, a sea anemone, a crab, or a mantis shrimp. Nobody knows how many other disguises it's capable of.

Of all the beautiful places Nastia and I have seen together, Lembeh Strait is the one we'd most like to return to. We still mention it in our conversations as a kind of paradise lost. I was glad that we managed to get there for her last few days before flying home.

A couple days later Sarit and Alex returned from Malaysia, and we took a ferry to the Moluccas, a group of islands east from Sulawesi. These lands had once been known as the Spice Islands, the most coveted prize for Arab and European traders. Local sultanates were the world's only producers of nutmeg, mace, and cloves. But during the Napoleonic Wars the British managed to transport a few seedlings to their colonies, breaking the monopoly and crashing the prices. Nowadays the Malukas Islands (as they are officially called) are the quiet backwater of Indonesia. Unfortunately, the relatively sparse population still manages to clear the forests at a catastrophic rate.

We had time to explore only two islands, the small Ternate and the larger Halmahera, which on the map looks like a miniature version of

Sulawesi, also K-shaped. Halmahera has extensive mangroves remaining, and I finally saw and heard my first saltwater croc there. But I made more interesting observations on Ternate, which has virtually no mangroves.

Ternate is a large volcano rising from the sea. In 1840, an eruption destroyed all forests there, including mangroves. Soon the volcanic ash turned into fertile soil that attracted massive human immigration, so the mangroves never recovered. But local crocodiles survived. They live in Tolire Besar, an emerald crater lake about two hundred yards in diameter, surrounded by vertical cliffs.

There are fewer than ten crocodiles in the lake. The saltwater crocodile is a species of seashores and large rivers, so I was curious to see if the "songs" of Ternate crocs, stuck in a small lake, were different from those of salties elsewhere. I managed to hear only two of them, and both were headslaps with no roars, just as the majority of this species' "songs" elsewhere. The sample size is too small, but if there really is no difference, it probably means that changes in "songs" take more than 170 years to evolve. On the other hand, it's not clear in which direction those "songs" should evolve in a population inhabiting just one small lake. Perhaps eventually the crocs will switch to producing only infrasound: since it's obvious which animal is "singing" and where it's located, there's no need for location beacons such as headslaps.

This was our last trip together. Sarit and Alex continued exploring Sulawesi but I had to find a place where saltwater crocodiles would be more numerous and easy to observe. Northern Australia seemed like an obvious choice, but spending a long time there would be too expensive. By that time my fellowship money and savings were all used up. I chose New Guinea.

34
Horror Stories
Crocodylus porosus

A good hunter's mind is in his prey's head.
—Papuan proverb

I FLEW TO SORONG, A CITY ON BIRD'S HEAD PENINSULA in the Indonesian part of New Guinea. It was my only chance. Indonesian New Guinea doesn't have a continuous road network, so the only way to get to other towns would be to fly (I couldn't afford that) or take a boat (for which I'd have to wait for weeks).

Even from the air, New Guinea looks very different from the rest of Indonesia. There's no endless chain of towns and villages along the coast; the forest starts from the beaches and stretches, unbroken, to the distant mountains of the mysterious interior. You can get from the city to the forest in hours rather than days. The local people are mostly dark-skinned, curly-haired Melanesians. They seem to find English easier to learn than Malay peoples do, so even in street restaurants you often run into English-speakers.

New Guinea is the world's second largest island, and its flora and fauna are still poorly known. Countless undescribed species hide in its forests and mountains, and many areas have no humans at all. How wonderful it would be to spend a few years exploring it. But for the moment I had no money to do anything. The only thing I could afford was renting a bicycle with a motor attached.

Soon I found a nice mangrove lagoon near a small village, part of a huge estuary with lots of crocodiles. All I had to do was watch them and hope they'd be close to me when they "sang." At low tide I could walk around a bit. For the first few days the low tide occurred in the morning, just when I needed it. I eventually observed six headslaps by five crocodiles, and only one of these slaps was accompanied by a weak growl. A local schoolteacher named Benyamin showed me the favorite hangouts of large males and told me a few horror stories about the crocs.

A trail of compacted mud connected the village with a bunch of huts on the far side of the lagoon. At low tide it looked like a narrow causeway and was often used by people, pigs, and dogs. At high tide it turned into a knee-deep ford, and trying to use it would be suicidal. Some of the crocodiles in the lagoon were over fifteen feet long. Local people were careful, but accidents did happen. Benyamin said that at least four people had been taken in the past few years.

Just one week before my arrival, a few fishermen were walking at night on a different trail, more than a hundred feet from the water's edge, followed by a stray dog. Suddenly, the people heard a squeak, looked back, and saw a large croc disappear in the grass with the dog in its jaws. Apparently, the reptile was lying in ambush near the trail, let the people pass, and caught the dog behind their backs.

I spent a lot of time watching the crocs near the causeway trail and saw them try sneaking up on dogs twice and on a calf once. Their tactics seemed to depend on the prey. Dogs were approached underwater. Both times the crocodile made it to within five feet of the dog but then had to jump out of the water, since it was too shallow near the trail. It didn't work because the dogs were very alert and agile, but in both cases the croc's jaws snapped less than a foot from the panicked mutt. The calf was approached by a large crocodile that glided toward it with its eyes and nostrils above the water. The calf didn't show any sign of recognizing the danger. The crocodile would certainly have gotten it (crocodilians have the strongest bites among living animals, and the jaws of a large saltie are strong enough to crush the skull of an adult bull), but a teenage boy walking behind the calf noticed the croc and threw a stick at it. The crocodile sank and disappeared.

Then one evening I witnessed the most complex hunting behavior I've ever observed in six years of watching crocodilians.

A pig was walking along the trail very cautiously, glancing nervously at floating leaves and twigs. I was sitting on a tree branch ten feet above the water and could see a large crocodile moving under the surface to intercept the pig. The croc got to about fifteen feet from the pig and suddenly erupted from the water like a submarine-launched missile, jaws agape, tail splashing wildly. This looked more like a threat display than a predatory attack. The pig panicked (who wouldn't?) and rushed away from the croc, leaving the trail and wading into the water on the other side.

My eyes were on the crocodile, so I didn't notice the two others until the very last moment. They were hiding in the water on the other side of the trail; I'm pretty sure they took that position after seeing the approaching pig. The hapless mammal ran straight into them. The light was too poor to see the details of what happened next. There was a huge splash, and the pig vanished, never making a single sound. The first crocodile ran across the trail to join the other two, and they all disappeared in the black waters of the lagoon.

Wolves and lions use this chase-into-ambush technique frequently, but no zoologist has ever observed any reptile do this. Mine was a singular observation, so I couldn't publish it in a scientific journal. How could I be certain that the crocs were coordinating their actions? Maybe the two crocs on the other side of the trail just happened to be in the right place at the right time? I was sure the whole thing was well planned and perfectly executed, but I'd never be able to convince a reviewer or an editor. And if I tried to see more hunts like that, it could take decades.

A few years later a reader of my blog emailed me a scanned page from the diary of Nicholas Miklouho-Maclay, an intrepid nineteenth-century Russian anthropologist. Miklouho-Maclay was one of the first followers of Darwin but also the first scientist to recognize the danger of applying Darwin's theories to human politics. He spent years living among the natives of New Guinea, and his revolutionary research proved that all humans belonged to the same species. Prior to his work, the prevailing scientific view was that "coloreds" were intermediate forms between white people and apes.

In his diary Miklouho-Maclay described team fishing by saltwater crocodiles, shown to him by a Papuan hunter. A larger crocodile would chase fish toward the shore with powerful splashes of its tail, while smaller, more agile crocs would wait in the shallows, ready to snatch the cornered fish.

So it looks like salties really do chase their prey into ambushes, at least in New Guinea.

You'd expect such cunning predators to be feasting on people every day, emptying villages and provinces. But, as Brandon Sideleau's newspaper survey has shown, saltwater crocs kill only about a hundred people per year (and even that is a lot more than previously thought). One reason is that many crocodiles are still afraid of humans after decades of large-scale hunting; the other is that few people live in mangroves.

However, no crocodile horror story is too scary to be true. There are old tales of boats capsizing near mangrove-lined shores and their entire crews disappearing before making it to dry land. There's a report by Miklouho-Maclay of a stilt village in northeastern New Guinea being inundated up to house floors by a rainstorm. Crocodiles broke into houses and tore to pieces every single inhabitant, except for a child that hid in a basket. And during the Battle of Ramree Island off Burma in 1945, at least four hundred Japanese soldiers, all armed to the teeth, were killed by saltwater crocodiles in one moonless night as they tried to escape the advancing British Indian troops by hiding in a mangrove swamp.

Salties would sometimes attack scuba divers underwater. Nile crocs and American alligators are also known to do that, although such cases are extremely rare. I've met curious crocs and gators underwater a few times, and always made them turn away by swimming straight toward them. But I've never tried this with a croc more than fifteen feet long, and I hope I'll never have to.

Meanwhile, the strangest thing happened to me. I got sick. I almost never get seriously sick while traveling—the only exception in the last twenty years was that brief spell of malaria in Kenya. But when I was in the Moluccas, I noticed that a few scratches on my leg wouldn't heal. Soon those scratches turned into scary-looking sores, all my lymph nodes became swollen and hurt a lot, and I had a high fever and could barely walk. Amusingly, it was looking more and more like Black Death symptoms. I had left my medicine chest to Sarit and Alex because they were going to stay in Indonesia longer. The only antibiotic I had didn't help.

Fortunately, the manager of a dive resort in Raja Ampat Islands invited me to stay there for a few days if I'd write an article about it for a Russian dive magazine. Clean seawater made all my sores dry up and heal within three days, but so much tissue was already dead that deep holes remained in my leg. Four years later, the scars are still visible.

Raja Ampat Islands are mostly tall limestone outcrops jutting from the sea alone or in groups and surrounded by rings upon rings of coral reefs. From the air this landscape seems too beautiful to be real, like a computer-synthesized background for a sci-fi movie. The reefs of Raja Ampat are the most diverse marine ecosystem on Earth. You can easily see two hundred fish species in one dive. What an idiot I was to go there just for a few days, and without Nastia.

But I had to leave as soon as my leg healed. I hitched a boat ride to Waigeo, the largest island in the archipelago. Local fishermen took me to a small bay where a particularly big crocodile lived, and the next morning I recorded the loudest headslaps I'd ever heard. This croc was a real monster; its head was over two feet long, and its total length was probably something like twenty feet. Except for those headslaps, it didn't do much during the time I was observing it, but once I saw it snatch a five-inch mudskipper (a small terrestrial fish).

I took a ferry back to Sorong. Benyamin had promised to ask local hunters about possible places to look for the New Guinea crocodile. This is a separate species, smaller than the saltie (usually less than ten feet long) and confined to freshwater. It used to be common in the interior, but, being easier to hunt than its scary relative, it became very rare by the 1970s. Nowadays, it's said to be common again but apparently not in the Sorong area. Local hunters had no idea what Benyamin was talking about.

The New Guinea crocodile inhabits rivers, lakes, and swamps. My theory predicted that, as a habitat generalist, it should include both roars and headslaps into its "songs." Later I managed to find one old crocodile of this species in a zoo, watched it for a long time, and eventually saw one "song," which happened to be an infrasound pulse followed by a snort-like roar and a headslap. But one is a very small sample size.

There is a very closely related species in the Philippines that is also small and confined to freshwater. But the Philippine crocodile seldom lives in rivers; its habitat is small lakes and forest swamps. So my theory said that it should use few, if any, headslaps. Indeed, records by people who have managed to breed this critically endangered species in captivity mention no headslaps, only strange "high-pitched groaning and bellowing sounds." I'd love to hear these sounds myself, but so far I haven't had a chance to observe these crocs, except for one sighting on the island of Mindanao long before I got interested in croc "songs."

I spent a full day in an Internet café looking for the cheapest way home. As usual, the optimal route was totally illogical. I had to return to Sulawesi, fly to Brisbane in Australia, somehow get to Sydney, and take a flight to New York from there. As a bonus, it would give me ten days in Australia.

When I landed in Brisbane and rented a car, I realized that I had no money left for food. I still needed to pay for a lot of fuel, because I wanted first to drive north, to the crocodile country, and then back south to Sydney.

One of the greatest pleasures in the life of a naturalist is coming to a new continent, where almost all plants and animals are unfamiliar. For me, Australia was the last new continent, and the last chance to experience that feeling of entering a completely new world.

Australia is like Canada in the tropics: roads are good, people are calm and polite, and the interior is a beautiful wilderness. There are hundreds of nature reserves, and even in towns you can see plenty of wildlife: cockatoos, rainbow lorikeets, flying foxes, opossums, even an occasional koala or wallaby.

There are two species of crocodiles in Australia. The saltwater crocodile is more widespread. In the northern interior lives the much smaller Johnston's crocodile, commonly called the "freshwater croc." It's said to be able to gallop at twenty miles per hour, faster than any other crocodilian.

Both species have been studied a lot in Australia, so I already knew from literature that Johnston's crocs, which live in rivers, lakes, and swamps, are more vocal than salties. Their "songs" include loud grunts as well as headslaps. It would take me almost a week to reach them in far northern Australia. Besides, it wasn't their mating season.

The only practical thing I could do was to try to hear at least one growl of an Australian saltie and check if it sounds the same as the growls of saltwater crocs in New Guinea and the Moluccas. I drove for two full days to get to Conway National Park, the closest place with a large saltie population, but found only three crocs there, and they didn't "sing." It was probably too cold for them. I was later told by Australian zoologists that I was a few weeks too early.

So I turned back south and drove to Sydney, stopping to see dancing lyrebirds, flowering *Banksia* trees, and other winter wonders. The beaches were already filling up with shorebirds. These brave creatures fly to Australia every year from the Arctic tundra, sometimes crossing thousands of miles of the stormy Pacific. Seeing them always makes me dream of the North. I like it so much, but working with crocodiles made it difficult for me to go there often. Driving through Australia, I was also thinking about the spoon-billed sandpiper of Chukotka, the bird so special to me. Every September, when ornithological expeditions return from the field, I anxiously await the results of the most recent nest counts, worrying that there might be none left.

In Sydney I stayed with a friend I'd met only on the Internet before. His name was Casey. He was a brilliant physicist, and we talked a lot about infrasound and other aspects of crocodilian communication. His initial calculations suggested that in order to produce infrasound underwater, an animal has to be at least two hundred feet long. Later he adjusted his estimate to sixty feet. That explains why the only other animals known to produce infrasound underwater, the large baleen whales, can do it. But crocodilian infrasound remains a mystery.

It took me less than sixty hours to get from Sydney to Knoxville via Beijing and New York. It was mid-September, the leaves were just beginning to turn yellow, but it was cold and raining as if in late fall. The Chinese bus dropped me off in a dark parking lot; I looked around and saw Nastia's small silhouette under a streetlight. I've never been so happy to see anyone in my life.

35
The Wrong Flood
Caiman latirostris

The real voyage of discovery consists not in seeking new landscapes but in having new eyes.
—Marcel Proust

My life returned to a more regular pattern. I had to teach twice a week but tried to spend at least every other weekend with Nastia. The drive from Miami to Knoxville took about fifteen hours one way, and I soon learned which gas stations along the route had the best coffee, and which hotels had free Wi-Fi in their parking lots.

Southern Florida remained warm and interesting even in November. Huge cane toads greeted me with their endless trills as I walked at night from my car to my office. I had to park almost a mile away. Technically, there was free parking on residential streets nearby, but that was an expensive neighborhood, and local residents didn't like seeing my old car near their homes. If I left it there, they would sometimes steal the license plate and call the police to have the car towed. They didn't care that I had to pay hundreds of dollars to get it back.

I did my best to make my office more comfortable, but it still looked like an interrogation room. The main problem was having no shower. It took half an hour and some acrobatics to wash myself part by part in the lab sink. The only shower in the building was an emergency sprinkler for washing your eyes if you accidentally got acid in them. It was in the corridor outside the chemistry lab. I could use it late at night to wash myself above the waist but couldn't risk undressing completely because a few people in the department had a habit of working late. I suspected that a couple other students also lived in their offices, but nobody ever talked about it because nobody knew for sure if it was allowed.

I had a lot of work to do: teaching, sorting the data I'd recorded, and writing the African guidebook. Sarit and Alex went to Moscow to sign the contract with a publisher. I didn't have time to go with them, and they were unfamiliar with the Russian business world, so the contract they signed was a total rip-off and could leave us without legal rights to the book if the publisher decided to use any of the loopholes. A few months later the chief editor of the publishing house resigned, and the new one refused to publish the book because he considered our descriptions of popular African scams to be racist. That was ironic. A popular theme in modern Russian humor is making fun of the American concept of political correctness, but in reality political correctness becomes absurd in Russia more often than in the US. I found another publisher through my blog, but there was another delay because tour companies tried to block the publication so that people wouldn't consider traveling to Africa independently. The book eventually took three years to get published.

From Moscow, Sarit and Alex went to Kenya and found jobs in a dive center. To my surprise, they settled in Africa permanently and don't travel anymore. I wouldn't have minded moving to the tropics myself, but I had a thesis to complete and a family life to begin.

Nastia and I decided to get officially married that November but to postpone the wedding celebration until a less depressing time of year. We were both atheists, so the official part of getting married was simple. We went to the city hall, got a marriage license, and asked the clerk to perform the ceremony. He was a young guy, shy and inexperienced, but obviously happy for us. I expected the ceremony to be purely bureaucratic and boring, but somehow it was rather touching.

Back in Miami, I spent day after day subjecting my data on Nile crocodiles and American alligators to various statistical tests. My advisor, Steve, was really good at statistics and drilled me relentlessly on all aspects of applying mathematics to observational data. He also spent a lot of time making sure there were no gaps in my work that needed to be filled. One problem he pointed out was the absence of the so-called "interobserver reliability tests."

If you are using statistics on data gathered by more than one person, subtle differences between people in interpreting the observations might cause errors in the results. The interobserver reliability test determines if different people interpret their observations in the same way. You do it by having a few observers look at the same events and comparing their records.

There were two sets of observations that I had to check for interobserver reliability. One was the "songs" of Nile crocodiles. I had to make sure that there weren't too many errors in scoring them as containing or not containing roars. In some populations these roars were quiet, cough-like, and difficult to hear. The other issue concerned my method of recognizing individual yacare caimans by the dark spots on their jaws. I used that method two years earlier but never checked how many mistakes were made.

I couldn't afford another trip to Africa, but I knew that there were crocodile farms in Israel with lots of Nile crocs. Among the readers of my blog were a few former Soviet zoologists living in Israel, and two of them agreed to spend a few weekend mornings at one of those farms scoring "songs" of captive Nile crocs. These crocodiles were from Lake Victoria, so their "coughs" were very quiet, but my friends had no difficulty noticing them. In all cases their records were identical.

As for yacare caimans, I had nobody in their South American range I could ask for such a favor. My friend Paolo had moved from Brazil to Uruguay; he never told me why, but I suspected it had something to do with the husband of one of his countless female acquaintances.

I asked Nastia if she'd like to spend our winter break in Brazil. I thought a month of chasing caimans through flooded savannas and rain forests would make a perfect honeymoon. Besides, Brazil has six species of caimans, on some of which I had little data. There was one problem: most of these species lived in the north of the country, while the only place we could fly to was São Paulo in the south. Airfares to all other destinations in Brazil were too high, even to Manaus where our plane to São Paulo was making a landing. But I remembered that much of Brazil had an excellent road network, so I thought we could simply drive to the Amazon.

Every time I arrive in Brazil, I'm amazed at how rapidly this country is changing. It's living proof that a left-wing president can be better for the economy than most right-wing ones. Unfortunately, the cost of living changes, too. Travel in Brazil now costs about the same as in the US.

We rented a tiny car and drove north. The coast of southern Brazil is gorgeous. Granite mountains rise above endless beaches and secluded coves. The slopes are covered with the so-called Atlantic Forest, one of the most beautiful rain forests in the world. During the rainy season almost every tree there is covered with flowers. It was raining almost nonstop on the coast, but we still enjoyed it.

The most colorful city in Brazil is Salvador, where European architecture blends with African lightheartedness. We happened to get there on Christmas Eve and walked into the main cathedral just when the lives of the Holy Family were being reenacted. To our surprise, the plot had been modified to resemble a soap opera. After each chapter, the choir filled the podium. But it wasn't your regular choir: men were muscular mulattoes naked to the waist, and girls were wearing short transparent dresses. Each song was followed by an openly erotic dance. It was the best performance we've ever seen in a house of worship.

We turned inland, into *cerrado* savannas and then into arid *caatinga* with cacti and acacias. This was the poorest part of the country, but it was also changing for the better. In just two years since my previous visit, small towns had turned from slums into clean, neat settlements. Peasants weren't wearing rags anymore and looked like respectable farmers. Everybody was driving new cars, but roads were still being paved. Life here was much cheaper; for a dollar you could buy a few pounds of mangoes, or try one of twelve different kinds of juice in roadside *fruterias.*

We turned west, trying to get to the Amazon rain forests. But we couldn't reach them. For hundreds upon hundreds of miles, we were crossing fields, pastures, and shrubs growing where the primary forest had been destroyed. The entire southeastern portion of the planet's greatest forest had vanished.

The civilization along our route seemed to be slowly vanishing, too. This was the true backcountry, never mentioned in guidebooks or shown accurately on maps. There were now ferries instead of bridges; towns became a bit rundown and unreasonably expensive. Our main delight, the fantastic *fruterias,* disappeared as well. The best places to stay were love motels, usually located in the outskirts of towns. They are somewhat shady places where teenagers or people having an affair can meet without being seen. But they are cheap, secluded, and rooms there are always clean and comfortable.

Roads supposed to be paved highways were dirt tracks. Some didn't exist at all. Instead of gas stations, there were private houses, where some old guy would sell you a few liters of alcohol-gasoline mix if you were lucky to find him at home. Once we had to camp in front of such a house because the owner had gone to visit his relatives and didn't return until the next morning.

We realized that we didn't have enough time to get to the Amazon. So we turned southwest, into the vast Mato Grosso Plateau, and desperately tried to reach the upper Xingu River, one of the southernmost areas of pristine rain forest. But the roads got so bad that we were hardly able to move. Our car was very small, so we could push it out of mud if it got stuck, but there was no way it could make it through hundreds of miles of sticky clay that looked as though it had been plowed. Once we hit a stretch so slippery that our car skidded off the road and hit a tree while we were driving at two miles per hour. We were tired of fixing bumpers, tires, and wheel rims and were afraid of damaging the car more seriously. So we turned south without ever seeing any rain forest or any caimans. It took us three more days to return to civilization and pavement. We were now fairly close to the Pantanal, and headed toward it.

I expected to find the Pantanal flooded, but it was very dry: that year all the rain was falling along the coast. Wildlife was crowded around the remaining pools; some small lakes were literally filled with caimans. We stopped at a private ranch called Fazenda Santa Clara, a little paradise with tame peccaries, raccoons, crab-eating foxes, capybaras, and hyacinth macaws running or flying free around the buildings. We rented two horses, camped near a pond with a few hundred caimans, and quickly did our interobserver reliability test. I was happy to see that my method of recognizing individual caimans was completely reliable and easy to use, even for Nastia, who had never seen a yacare before.

After spending a few days in the Pantanal we drove south again. I was now looking for another species, the broad-snouted caiman. It's a small (usually five to eight feet long), funny-looking creature with the broadest head of any living crocodilian. Its broad jaws are adapted to cracking snail and turtle shells. Hunting these caimans has lead to outbreaks of parasitic worms in cattle because the larvae of those worms live in freshwater snails that caimans feed on. As in all crocodilians, the sex of hatchling broad-snouted caimans is determined by temperature. Some observations suggest that females of this species have learned to ensure that both boys and girls are born by laying eggs in two layers inside the nest. The bottom layer supposedly develops at a slightly higher temperature.

Except for broader heads, these caimans are very similar to yacares. Where the two species occur together, broad-snouted caimans prefer more marshy habitats, but they are habitat generalists and can be found anywhere from mangroves to cattle ponds. They apparently avoid interbreeding with yacares by having their mating season about two months later.

Broad-snouted caimans are widespread in coastal Brazil, but somehow we missed them when we were traveling along the coast. I finally found some at Itaipu Reservoir on the Brazil-Paraguay border.

The reservoir has formed above the world's largest hydroelectric dam. Submerged by its waters are the remains of Guaíra Falls, once the world's largest waterfalls by volume. They were dynamited by Brazil in the 1980s.

The "songs" of broad-snouted caimans seemed identical to those of yacares, so after recording a few I decided not to spend too much effort studying them. Instead, we drove to the famous Iguazu Falls.

I first visited Iguazu Falls in 1995. At that time I'd been hitchhiking around Latin America for almost half a year and was trying to get back to the northern part of the continent after traveling in Patagonia. But obtaining a Brazilian visa took a lot of time. I had an invitation from my Brazilian friend Paolo, but the clerks at the consulate were so scared of visitors from an ex-Communist country that they refused to issue a visa. I had to camp in the forest on the Argentine side of the falls and go to the nearby town every morning to pay yet another visit to the consulate. After about a week I realized that only a bribe would help. That drained my resources so much that I couldn't complete the journey, and after a month in Brazil I had to fly home from São Paulo.

At that time the waterfalls were pristine, and visiting them wasn't regulated much. You could wait until the crowds were gone in the evening and have the views all to yourself. Lunar rainbows were particularly beautiful. You could also walk the forest trails freely, enjoying birds, butterflies, deer, and sometimes even jaguars.

Things have changed. So many tourists flock to the falls nowadays that visits are tightly regulated, and queues for entry tickets are hours long. What is even worse is that you aren't allowed to be at the falls at night unless you stay at an insanely expensive hotel, where you can arrange for the guards to escort you two hundred yards to the viewpoints. All other people can see the place only in the midday heat, when there's no wildlife around except for garbage-eating coatis. The extensive forests around the falls are also off-limits; the only way to see them is by taking a guided walk along a single trail, which costs about a hundred dollars. Even the views have been spoiled. An outstandingly ugly Sheraton building now sits on top of the falls on the Argentine side. One of the world's most beautiful places has been turned into a theme park.

We returned to the coast, where it was still raining. Hundreds of people in Brazil were killed by floods and landslides that winter. We stopped on a small offshore island for a couple days to rest from a month of almost nonstop driving and returned to São Paulo. We arrived there late at night during a rainstorm that dumped more than twenty inches of rain in two hours and turned most streets into torrents. We found a store with two hundred kinds of cakes and pastries, waited out the storm, drove to a love motel, slept for a few hours, straightened a few dents in the radiator and the hood cover, and went to return the car in the morning. Until the last moment I was afraid that our brave little vehicle would fall apart in the middle of the street after all the bumps, potholes, and stream crossings it had been through. But everything worked out fine. The rental company later tried to charge our credit card twice, but—surprise!—we noticed the extra $1,400 charge and got it refunded.

We thought our adventures were over when we got on a plane to Miami, but they weren't.

36
Unsolved Riddles
Alligator mississippiensis

The love for all living creatures is the noblest attribute of man.
—Charles Darwin

Our plan was to spend the night in my office and fly home from Fort Lauderdale the next day. We had no winter clothes and were surprised to see ice on puddles when we landed in Miami. We almost froze to death waiting for a bus to the university. Southern Florida was in the middle of its worst freeze in decades. It lasted for almost a week.

I left Nastia in Knoxville and immediately drove back to Florida to see how the extreme weather affected alligators and crocodiles. By that time the freeze was over, but it was still unusually cold and it didn't get warm until March. Air temperatures in southern Florida drop below freezing almost every year, but this was the first time in decades when the cold temperatures lasted long enough to cool down the sea. Mangrove channels and lagoons were full of dead fish. Hundreds of manatees and sea turtles perished. As for the terrestrial wildlife, native Florida species survived the freeze pretty well, but many introduced reptiles such as Burmese pythons were almost wiped out.

Alligators didn't seem to suffer much, but dozens of crocodiles didn't make it. Those that survived started mating a month later than usual. Introduced spectacled caimans fared even worse; all populations except the southernmost one died out.

I started putting together the first fragments of my dissertation text—slow, meticulous work. I remembered Steve saying three years earlier that my project was too big and complex for a dissertation study. I thought he was concerned with the amount of fieldwork, but he was also talking about the write-up.

I still couldn't explain the strange results of my alligator and crocodile studies. Why did one of the two key predictions of my theory work only for alligators and the other only for crocodiles? All alligators bellowed a lot, but only those living in big lakes and rivers headslapped frequently. With Nile crocodiles it was the opposite: they all headslapped a lot, but only those in small ponds roared often and had loud roars.

As soon as I had these results in readable form, I sent them to John Thorbjarnarson. If anyone could explain this mystery, it was him. Over the years he had helped me a great deal, and I couldn't wait to know his opinion. But, for the first time ever, he didn't answer my email. I thought he was somewhere in the field, without Internet access.

And then I received an email from one of John's colleagues. John was dead. He died from cerebral malaria a few days after flying from Africa to India. Nobody knew exactly why it happened. Perhaps Indian doctors used antimalarial drugs for which African strains of malaria had already evolved resistance. Or maybe John was so busy that he didn't start taking medicine soon enough.

He was only fifty-two and had spent much of his life in tropical swamps, often working at night, when malaria-carrying mosquitoes were the most active. He was certainly aware of the risk.

It was a tragic loss. John had done more than anyone for saving endangered crocodilians. Two or three species would probably be extinct without his work. His openness and humor helped him succeed in launching conservation programs in places where people were initially very hostile to the idea. He helped dozens of young biologists and conservationists. As someone said at his memorial service, "Few people have so many friends and not a single enemy."

The next alligator mating season was my last chance to solve the puzzle. It was my fifth year at the university; I was expected to defend my dissertation no later than in April of the sixth year.

I had to somehow improve my data analysis. Steve pointed out that I couldn't use some powerful statistical tests because my alligator data was "unbalanced," meaning that my selection of study sites was asymmetrical. My original plan was to divide the range of the American alligator into four quadrants, and to have one "big lake" and one "small pond" study site in each quadrant. But populations living only in small ponds proved difficult to find. I had data from only two, one in the southeastern quadrant and one in the northwest. I also had three "big lake" sites, one in each quadrant except the southeast.

I couldn't find more "small pond" sites. But if I could add one "big lake" population in the southeastern quadrant, my data would become partially balanced: if I divided the alligator's range in two halves (northern and southern or eastern and western), I'd have one "small pond" population and two "big lake" populations in each half. So I could compare similar sets of data, and get more clear-cut results.

There were many places in Florida where alligators lived only in big lakes and lagoons. The easiest to work in was Merritt Island National Wildlife Refuge, just north of Cape Canaveral. I went there in late March and easily found a few gators, but there was no sign of any mating-related activity yet.

There was one other population of American alligators that I wanted to study, even though it couldn't help me with balancing my design. It inhabited a few islands in the central part of the Florida Keys. The Keys have the warmest climate in the continental US, so alligators there mate earlier than elsewhere. I decided to spend some time there until the Merritt Island gators would begin "singing."

The Florida Keys are a long island chain composed of sand and limestone. The central part of the chain has the largest islands, with extensive pine forests and a few freshwater lakes in sinkholes. This area is famous for the dwarf race of white-tailed deer, about half the regular size. Other unique inhabitants of the Keys include the local race of marsh rabbit, named after Hugh Hefner, the founder of *Playboy*, who had financed studying it. All these animals most likely reached the Keys during the last Ice Age, when the sea levels were lower and the islands were all connected to the mainland.

There are also alligators on the central Keys. They are smaller than those on the mainland but otherwise seem identical, and it is unknown if they are Ice Age relics or recent immigrants. I think the latter is more likely. The Keys population is probably too small to survive for thousands of years in full isolation, and crossing to the Keys is not that difficult for alligators. Even though they generally don't like saltwater, they've been seen at sea twenty miles offshore.

There are probably between twenty and forty adult alligators in the Keys. They inhabit tiny forest sinkholes and can probably swim between them through underground channels. Migrating overland is difficult nowadays, because much of the Keys have been privatized, subdivided, and fenced.

The trickiest part of my work in the Keys was finding the sinkholes. They were too small to see in Google Earth, so I had to ask locals if they had any on their property. People of the remote Keys backwoods are a colorful bunch, with ultra-right-wing survivalists living side by side with ex-hippie tree huggers. Most of them are there because they don't like strangers, and some are a bit paranoid, especially about people with a Russian accent, secret agents searching for endangered species as a pretext to restricting private landowners' rights, or anyone looking for alligators of which the Keys residents tend to be rather fond.

The sinkholes were so small that each had just one alligator, if any. In early April adult gators began moving from their sinkholes to small artificial lakes. There were two such lakes on the islands, and one of them soon had a decent bellowing chorus. Alligators in the Keys bellowed just like those on the mainland, but I never saw them headslapping.

I observed that lake from a short boardwalk popular with tourists. One morning there was a family with a boy of about five. He was fascinated with alligators and watched them so intensely that he stuck his head out between the bars of the handrail. Two larger gators immediately started floating toward him, smiling gently, ready to take care of the child if he accidentally fell into the water. "Daddy, look!"

the kid shouted. "Two gators are coming here! They are attracted to something! What do you think it could be?"

On another occasion, a woman got so excited from seeing her first alligators that she accidentally dropped an expensive-looking ring from her finger into the lake, and then went almost hysterical. I told her I could get it out (it was in a foot of water, and the alligators in the lake weren't big enough to be seriously dangerous for an adult), but at that moment a gator picked the ring up from the bottom and swallowed it. Everybody fell silent for a minute, then the woman asked: "Could you please get it out of the lake when it comes out?"

I explained that the ring was unlikely to come out any time soon. Crocodilians (and birds) often swallow pebbles and other hard objects that accumulate in their stomachs. Neither crocodilians nor birds can chew, so they use the so-called *gastric mill*: their muscular stomachs contract periodically and the pebbles gradually crush the food. Researchers sometimes track crocodiles by making them swallow small radio transmitters in bulletproof casings.

I kept thinking about those differences between alligators and crocodiles. Why did all alligators bellow a lot, while some crocodiles hardly ever roared? Could some feature of alligator biology require them to bellow even in large lakes and rivers, where infrasound and headslaps could carry much farther?

One thing that gators have and crocs don't is the group courtship: bellowing choruses in the morning and "dances" at night. What if alligator bellows aren't just individual calls inviting members of the opposite sex to visit that particular gator? What if they have a second function? Maybe the purpose of bellowing choruses is to gather as many animals as possible at a particular location, so that everybody would have a good choice of potential mates? And headslaps can't serve that second function, because it's difficult to tell how many animals are headslapping.

Crocodiles don't have group courtship, so their roars are simply individual "songs." That's why they are used only in places where they carry information better than headslaps and infrasound.

I thought I found an answer to half of the riddle. But why do all crocodiles use a lot of headslaps? Maybe headslaps have some second function in the "songs" of crocodiles but not of alligators. What could it be? I didn't know.

In mid-April the nights finally got warm, and alligators at Merritt Island began to "sing." I was done with teaching for that semester, so I could stay in the field all the time, eating clams and sleeping in the car.

Alligators at Merritt Island lived in brackish marshes. The water was so murky that they probably couldn't see anything in it, but they looked reasonably well fed. Alligators and caimans have little black pits on their jaws, called dermal pressure receptors. These tiny organs can sense pressure changes and allow the animals to catch fish without seeing them. Crocodiles have such pits not only on their jaws but also on the body scales. The pits are sensory organs, reacting to vibrations, temperature changes, and chemical clues. They are an interesting engineering solution, making the skin sensitive and armored at the same time.

At first working at Merritt Island was easy. I found a few large males and recorded their "songs." But the road network at the refuge was limited, and I soon ran out of gators in accessible places, so I had to explore every jeep track in the marshes.

One night, as I was slowly driving along a familiar dirt road on the lagoon shore, the dry crust suddenly gave way, and my car sank into black stinky mud full of oyster shards.

I spent four hours digging a deep trench to drain the water from the puddle into the lagoon, then tried to free the car. But the sticky mud was holding it so tightly that I couldn't jack it up to put sticks under the wheels. My shovel wouldn't fit under the car, so I had to dig with my hands.

A large alligator floating nearby was very sympathetic to my troubles. Every time I went to the shore to wash mud off my hands and legs, he'd slowly approach me with a shy, friendly smile, as if offering to help. I considered using him to pull the car out (alligators are much stronger than they look), but I didn't have a long rope.

The worst thing was, every time I washed the mud off my hands and legs I also washed off the bug spray. Soon I was all polka-dotted with itchy no-see-um bites.

After midnight it became clear that I wouldn't be able to free the car. Luckily, I didn't have to walk out: my cell phone worked there. I called 911, they forwarded my call to Kennedy Space Center security, and in half an hour I was joined by three security guards and a policeman. They arrived in four huge SUVs. I was sure it would take them two minutes to pull my car out.

But my joy was premature. First they interrogated me for an hour. They even googled me (every SUV had satellite Internet), found my website, and asked me a lot of questions about it. The policeman couldn't believe that I dug the trench by myself, and kept accusing me of making up the whole story. What saved me from a trip to the county jail was that three years earlier I had obtained a permit to enter the Kennedy Space Center to help a friend of mine with his beach mice research and to take Kami to the beautiful pristine beaches there.

Then the security officers informed me that they couldn't use government vehicles to pull me out. I had to call a tow truck from the nearby town. Another hour later, an old guy in a tow truck showed up, chatted with me for five minutes about alligators, pulled the car out, charged me a hundred dollars, and left.

The security guards left, too, but the policeman didn't. He told me the reason I got stuck was that I took a wrong route across the puddle. I disagreed. He jumped in his SUV to show me where the correct route was and promptly got stuck.

He called the security guards to pull him out. They returned, laughed at him, and said, "Sorry, bro, can't do nothing in 'em government vehicles!" By the time I had washed myself in the bay and was leaving, he was calling back the tow truck.

My hands were all cut by oyster shards, but it was a great morning nonetheless. At ten o'clock I recorded the very last gator "song" I needed. I cleaned the car wheels of mud and drove straight to Knoxville. I had just one week left before my wedding celebration.

37

Finding Answers

Caiman crocodilus

To see what is in front of one's nose requires a constant struggle.
—George Orwell

I WAS FALLING IN LOVE WITH MY WIFE. It was a bit strange after having been together for more than a year, but we had spent most of that time apart, and every precious opportunity to see each other still felt like a date, like meeting a new person.

Our wedding was in mid-May. Unfortunately, most of the people we'd have liked to invite were scattered all over the world. We had only about twenty guests, but they had been born in ten countries. Our mothers came from Ukraine and Russia.

Since I was in Florida for most of the spring, Nastia did the planning. She designed the wedding in old Slavic style with lots of funny games and a colossal sweet bread called *karavai*, made by our friend, who spent a whole day baking it in accordance with an ancient fertility rite. We rented a cabin on a hilltop with great views of the Smokies. Mountain laurel was blooming all around us. But the nicest touch wasn't planned: during the feast, a bear showed up under our balcony, as if the forest was sending its greetings.

I knew that getting married would mean big changes in my life. Nastia liked traveling a lot, but she was a bit less adventurous than me.

She loved her work and enjoyed having a cozy home. Never again would I be able to wander around the world for many months at a time. It was a bigger sacrifice than you might think, because I enjoyed seeing new places more than anything. But it was worth it.

We tried to make the best of that summer because we knew it was going to be short. I had to be back in Florida by mid-August; that meant being separated again. All those long phone talks, trying to tell each other things that can be said only by touching. All those endless evenings, when we could see each other in video chats, but that didn't make the hundreds of miles between us shrink by an inch. All those nights spent together—but alone, in empty rooms lit with the cold blue lights of our computer screens.

But for now we were together. We spent a month driving around California and Oregon, then went to Nicaragua for a week of scuba diving in the Corn Islands. I halfheartedly considered looking for spectacled caimans at Lake Nicaragua but was almost happy to learn from local biologists that the mating season was more than a month away. In July we flew to Alaska and explored much of that state, from Nome, Barrow, and Deadhorse to Homer and Yakutat.

And then it was over. I had to return to Florida for another year of teaching and writing.

I already had more data than I'd ever hoped to gather. However, I still had a few bits of fieldwork to do. One species I still hadn't observed "singing" was the spectacled caiman. It's the most widespread crocodilian in the New World, occurring from Mexico to the Amazon. Despite having poor-quality skin, it's hunted heavily throughout much of its range, but it's probably more common now than in the past, because its larger competitors—American and Orinoco crocodiles and the black caiman—have been hunted out in many places. Over a million spectacled caiman skins are produced annually; that's more than half of all crocodilian skins entering the market.

I'd seen lots of these caimans in Guatemala, Venezuela, Colombia, Guyana, Trinidad, and Brazil, but every time the timing wasn't right and they didn't "sing." At first it didn't bother me much, because I knew I could always find them, and besides, there were detailed descriptions of their "songs" in literature. But now I thought it would be a shame to have only secondhand data on the most common crocodilian in the Americas. And I didn't have to go very far to observe it.

In the 1960s, baby caimans were popular in the pet trade, and thousands were imported from Colombia and Central America to the US. As they became too big to keep in fish tanks, many were released into the wild. The only place warm enough for them to survive was Florida, where they joined dozens of other exotic species wreaking havoc on fragile native ecosystems.

Florida is so full of introduced species that in some places you don't see many native ones. When I walked around our department building, almost all the animals I saw were nonnatives: anole lizards from various Caribbean islands, Cuban tree frogs and green iguanas, Mediterranean geckos, Asian blind snakes and tiger mosquitoes, South American cane toads and parakeets. There was even a flock of blue-and-yellow macaws feeding in palm trees; amazingly, people walking below hardly ever noticed these huge, noisy birds flying thirty feet above their heads.

The exotic pet trade is almost as lucrative a business as smuggling drugs, but what few people realize is that it's also every bit as destructive. Just in southern Florida, it has been responsible for introducing cobras, Burmese pythons, Nile monitors, and tegu lizards—all of them large, highly effective predators. But what is even worse is that the exotic pet trade spreads diseases, from monkeypox to numerous pathogens unknown to science. Trade in frogs led to the hybridization of two strains of frog-infesting chytrid fungus; the resulting super-fungus has escaped into the wild and caused total extinction of hundreds of amphibian species on all continents.

Fortunately, spectacled caimans in Florida didn't become the disaster they could have. The first introduced population lived on the grounds of Homestead Air Reserve Base. The Air Reserve took it upon itself to get rid of these unwelcome immigrants and over the years managed to kill almost all of them. Populations farther north perished in the big freeze of 2010.

In the 1990s the Air Reserve Base was partially closed and much of its territory was opened to development. The development was slow to come, so there are lots of narrow streets leading nowhere, with overgrown shoulders and grass breaking through the pavement.

Caimans used to live in drainage canals there, but all of them had been killed, or at least I couldn't find any. There were only alligators. However, during my first year of studying crocodiles I found a small lake in an abandoned amusement park that had three caimans in it. The lake was fenced off and surrounded with dense shrubbery, and when I found it, the caimans were still juveniles. Now, four years later, I expected them to be adults.

I drove to that lake and was surprised to see a new sign saying, NO TRESPASSING—DANGEROUS WILDLIFE. I immediately climbed over the fence and searched the lakeshores but found nothing dangerous except voracious mosquitoes. I still have no idea what the sign was about or who had posted it.

There was only one caiman left in the lake. It was a male about five feet long, and it "sang" a lot. In two very hot and humid days I recorded six roaring displays and two headslapping ones. The "songs" were identical to those of yacare caimans. I decided there was no point in continuing.

The next three months were very busy. Writing even a short scientific paper is a lot of work, and my thesis was a combination of four large papers plus a few smaller ones. The chapter that was giving me the most trouble was the overview of "songs" of all crocodilian species. One problem was that the data I had for different species was difficult to compare: for some I had a lot of information, for others just a single observation or some vague descriptions from literature. I also had to make sure that I hadn't missed any publications on the subject, so I had to go through hundreds of papers and books on crocodilian biology. Old books and obscure journals published in Africa, India, or Latin America were difficult to find. Of course, if I were to have done this in the pre-Internet era, or in some less developed country, my task would be close to impossible. But in the US I could enjoy the eighth wonder of the world, the university library system with its interlibrary loan service. Sometimes I had to wait for two to three months for some particularly exotic edition, but eventually I received everything I asked for. And most of my colleagues took that miracle for granted.

I also kept writing to crocodile researchers all over the world. I met many of them on an Internet Listserv called *Croclist.* Since there weren't that many crocodilian researchers, this forum was not a particularly busy one—though sometimes you could learn interesting things there, or have a good laugh. Once a child asked: "What is it that crocodiles can do and other animals can't?" The scientific community took the question seriously. The suggested answers included fasting for a year, communicating with sound and infrasound at the same time, and dealing in a unique way with cuts and other injuries: the crocodiles' bodies quickly produce a lot of collagen fiber, blocking off the wound and any possible infection (croc blood also contains a powerful antibiotic called crocodillin). But the boy was kind enough to provide his own answer, which was much simpler: "They can make baby crocodiles!"

One day I received an email from Soham Mukherjee, a young Indian zoologist working in the Madras Crocodile Bank. They had a few African narrow-snouted crocodiles there. Soham told me that these crocs were both roaring and headslapping. Then he said something that caught my attention. "We have a pool with a few large females," he wrote, "and they are headslapping a lot."

That was strange. I'd never seen a female crocodile headslap. I asked Soham if there were any males in that pool. "No," he answered, "just large females, and the largest one headslaps all the time."

OK, I thought, so there's no dominant male, the largest female is the dominant animal, and that's why . . . Of course, headslapping is a signal of dominance. Hadn't I already read it somewhere?

I ran through a few old books. Indeed, it was a well-known fact. Headslapping was a signal of dominance in crocodiles. But that was also the answer to the second half of my crocs versus gators mystery! It was under my nose the whole time! How could I have been so stupid?

Now all the pieces of the puzzle were in place. My theory predicted that both alligators and crocodiles should bellow or roar more often in small pools but headslap more often in big lakes and rivers. It works for headslaps in alligators, and for roars in crocodiles.

But it doesn't work for bellows in alligators: all alligators bellow a lot because their bellows have a second function, attracting more animals to the sites of nighttime "dances." And it doesn't work for headslaps in crocodiles, because headslaps in crocodiles have a second function, being signals of dominance. In alligators headslaps don't have that second function, because alligators aren't so territorial and don't have well-established hierarchies.

It was beautiful. Now I knew I was going to have a good dissertation. But there was still a lot of work to do.

38
Through the Mist
Crocodylus acutus

Why would God create these forests if not for us to find gold in them?
—Francisco de Orellana

We decided to spend our winter break in Ecuador. I expected American crocodiles and some caimans to "sing" there. Our plan was to fly to Quito, study crocs at the coast, then cross the Andes and look for caimans in the Amazon. This time the cheapest flights happened to be from New York, so we first drove there.

Quito is located in a broad valley between two parallel rows of snow-clad volcanoes. It was the rainy season, so the mountains weren't visible, but the views from the hills above the city were still splendid. Sometimes a narrow ray of sunlight would squeeze through matching holes in cloud layers and make a palace, a cathedral, or a city block sparkle like a gem.

We rented a car and drove to the ocean. Much of coastal Ecuador is affected by a cold current. In northern Chile and Peru that current is cold enough to turn the entire coast into a desert, but in Ecuador it only makes the seaboard relatively dry and the skies gray. The cold current didn't prevent American crocodiles from colonizing that coast. They avoid the cold seawater by living in river estuaries. Almost all of them have been hunted out, with only two small populations remaining, one in northwestern Peru and one in Ecuador.

We stopped at a little *finca*, or homestead, owned by an Italian downshifter who wasn't familiar with that term because he'd been in Ecuador for over a decade. As with many coastal farms and plantations, the finca doubled as a makeshift ecolodge, mostly catering to weekend visitors from Quito. We rented a cabin in the garden, and at night we could hear exotic fruit falling on the roof, merry calls of geckos, and solemn hooting of small owls over the near-constant drumming of rain. Because of the rains, our host had to dry his harvest of cacao indoors. Drinking a lot of cheap cacao made us sleepy, and we slept through much of those rainy days but had to get up before sunrise to catch the crocodile "songs." We had to walk for half a mile and ford a river to get to our car; we couldn't park it closer to the finca because it could get stranded if the river rose too much. Even the road would often turn into a stream of red mud after a night of rain.

The big, misty estuary smelled of salt and seaweed. There were hundreds of small islands; some of them weren't attached to the bottom of the river, but slowly floated toward the sea after being torn from the forested shores upstream. These floating islands of trees interconnected by their roots are called *vegetation rafts*. They can drift across the ocean for hundreds of miles and are said to be the main source of nonflying animals found on remote islands, such as the Galápagos Islands.

Not having a boat, we had to wade through deep mud between mangrove roots to look for crocodiles. The few males that we found produced headslaps but virtually no roars. I think the ancestors of these crocs had all but dropped roars from their "songs" while slowly spreading south along the Pacific coast of South America, where they now live exclusively in large estuaries.

After a week of watching crocodiles, we headed inland. Crossing the Ecuadorean Andes during the rainy season means climbing endless switchbacks in dense fog to a frigid high pass where sparkling hummingbirds attend to strange flowers, then having a short break from the constant rains while crossing the Valley of Quito, ascending to a second high pass, and rolling down almost to sea level through more fog and increasingly warm rain. The densely forested mountain slopes are very scenic even in poor weather.

Andean forests are full of orchids, but usually only a tiny fraction of them is blooming at any given time. You can look through thousands of orchids decorating mossy tree branches and never find a single flower. But that year something unusual happened, probably thanks to the heavy rains. Almost all orchids were blooming, and the forests turned into spectacular orchid gardens, so moist that even the furniture in nature reserve offices smelled of wet moss. Every large tree branch had flowering orchids on it. They didn't look like the gaudy ones you see in flower shops. Wild mountain orchids are mostly small and nondescript, and even if they are everywhere, you still have to search for them. Some are so tiny that you need a magnifying glass to see the petals, others look like little bees or snails, yet others resemble dewdrops or bunches of dry leaves, and only a few are really colorful.

Soon the mountains ended sharply, and we saw the immense expanse of the Amazon lowlands, a sight that every naturalist cherishes above any other.

Ecuador owns a relatively small sector of the Amazon. The lands adjacent to the Andean foothills have mostly been deforested, but farther east, near the borders with Colombia and Peru, huge blocks of virtually untouched forest still exist. Oil companies are constantly pushing for more roads and pipelines and would completely destroy the forests in just a few years given a chance, but so far conservationists have been able to slow down the relentless advance of civilization.

The largest and most pristine forest is Yasuni National Park, one of the four places in South America claimed to have the world's highest biodiversity. Unfortunately, this region has large oil deposits. In 2007 President Rafael Correa threatened to open Yasuni for large-scale development unless his country was paid a $3.6 billion ransom—about half the price of the oil that supposedly could be extracted from Yasuni. But only about $300 million was raised. Most "rich" governments refused to pitch in, deeming the whole scheme a kind of blackmail and trying to avoid creating a precedent. So the money came from private donations, NGOs, communities, and even some countries where the standard of living is lower than in Ecuador. In August 2013 Correa declared the ransom insufficient and opened the park for drilling. Yasuni will now be turned into a wasteland of poisoned rivers and barren pastures, just like many parts of the Amazon that oil companies are already done with.

We arranged a short stay in Yasuni's only tourist lodge in exchange for promoting it on the Russian tourist market. The lodge was accessible from a small town called Coca, where the grounds of our hotel sheltered a lot of animals, mostly brought from distant forests by hunters and oil workers. There were squirrel monkeys, tiny black tamarins, agoutis always carrying mangoes in their mouths, colorful birds, and a baby acouchi (an animal similar to an agouti, but much smaller). The commotion never stopped: squirrel monkeys dropped tree fruit, tamarins dropped each other, agoutis quietly picked up the fruit, monkeys chased agoutis, macaws chased monkeys—it was amazing. Only the baby acouchi had nobody to quarrel with, so it was happy when I picked it up. It stomped around in my hands, peed in a friendly way, and ran on.

The boat ride to the lodge took four hours. Oil rigs, clear-cuts, and garbage piles lining the shores gave way to huge trees. We stopped in a small Indian village to change our fast motorized canoe for a small dugout with a teenage boatman, then left the turbid beige waters of the fast-flowing Rio Napo and continued up a narrow, quiet blackwater stream. I've been up many such streams in the Amazon, but I can never get used to their enchanting beauty, the stillness of dense forest, the sudden appearance of strange birds and monkeys in the trees, the silent murkiness of the water where countless bizarre creatures lurk, invisible. The stream kept getting narrower and darker, until suddenly the forest opened up. We were in a large lake, with the lodge on the far side, built on a small hill, almost an island—one of the most beautiful places I've seen in the Amazon, with great views of the lake and the forest, all the way to the Andes. The views were even better from a two-hundred-foot-tall canopy tower attached to a giant mahogany tree deep in the jungle.

The lodge was run by the villagers, Quechua Indians whose ancestors had come from the Andes centuries ago, trying to escape the conquistadores who were laying waste to the fallen Inca empire. Only some English-speaking guides were hired naturalists from Quito. Although there were about eighty adults (and as many children) in the village, the whole operation felt like a well-organized family business. We liked the place so much that later we spent a lot of time promoting it in Russia, eventually finding volunteers to teach local guides English and wealthy sponsors to contribute to ransoming Yasuni from Big Oil.

The lodge grounds were full of wildlife, with rare birds and monkeys coming from the forest for fruit and flowers. There was a network of foot and canoe trails where something interesting could be found literally every minute. One trail led to a clearwater stream with colorful fish, and then to a wooden blind overlooking a small cave where tiny parrots and huge scarlet macaws would come to dig for mineral salts. Another trail passed through a patch of gorgeous *Eucharis* flowers, which look like huge ghostly daffodils with a wonderful fragrance and grow almost exclusively in the rain forests of eastern Ecuador.

It was like a third honeymoon, even better than our first two. Rain forests of the Amazon are a well-known paradise for naturalists, anthropologists, and exotic food lovers. Much less known are the exquisite erotic pleasures you can enjoy here. For example, you can cover the body of your loved one with insect repellent, making sure not to miss a single square inch of her tan skin, then wash it off the next moring under a forest waterfall. You can bathe with your loved one in a round hollow left in the sandy bottom of a cool jungle stream by a large stingray and enjoy the soft lips of tiny fish kissing your most tender body parts, or stroke your loved one's skin with a finger covered with wild honey, so hundreds of colorful butterflies will suck the honey up with their soft proboscis. Or you can hug each other trying to get warm under plastic raincoats during a fierce rainstorm with lightning flashing ten times a second, thorny palms swinging widely in the wind, and orchids, torn from tree branches, falling in the water all around you.

Locals engage in far more hedonistic entertainment: smoking magic mushrooms of at least six kinds, using giant pulsating grubs as female sex toys, drinking the unusually stimulating juice of *Anthurium* berries. But we weren't sufficiently bored to try those.

There were two species of caimans around. Black caimans inhabited the lake and the stream connecting it with Rio Napo. I watched one large male. His "songs" were identical to black caiman displays I'd seen in Guyana, so I switched to searching for dwarf caimans.

That was trickier. Dwarf caimans lived in the smallest streams, too shallow for any canoe. I tried walking parallel to the streams, but the surrounding forest was flooded, so I was making too much noise.

I was about to give up when one night I found a clearwater creek flowing parallel to a low hill. Now I could move along the shore quietly and soon located a Schneider's dwarf caiman that hid in a little pool during the day and explored the flooded forest at night. I don't know what it hunted there, but once it coughed up a hair ball consisting of short fur, probably from some rodent. Crocodilians have highly concentrated gastric acid and easily digest bone, hooves, and tortoiseshell, but for some reason they can't digest hair, even though hair is made of the same kind of protein (called *keratin*) as hooves.

The little caiman "sang" occasionally, making either roaring or headslapping displays. There was nothing new about it, but the life of dwarf caimans in the wild was so little known that any bit of data was precious.

Serverio, the head guide at the lodge, was very knowledgeable about wildlife. He had no education; his father was the night shaman at the village (a night shaman is responsible for curing sick people, while a day shaman works mostly in prevention). But he knew the calls of hundreds of bird species and lots of interesting things about the forest. He said December wasn't the normal "singing" time for Yasuni caimans. He thought the change in their behavior was caused by the heavy rains. I don't know if it was true. Over the years I've heard other species, particularly alligators, "sing" out of season occasionally but never noticed any connection with the weather.

Having to leave the wonderful lodge after just a few days was downright painful. We returned to Coca and drove back to Quito. Only when the airplane rose above the clouds did we see the surrounding volcanoes in all their splendor. It was South America's farewell gift to us. We didn't know if we'd ever be able to return there. Our easy lives as graduate students were coming to an end.

39
The Last Song
Crocodylus siamensis

When you are everywhere, you are nowhere. When you are somewhere, you are everywhere.
—Rumi

YOU ALREADY KNOW THE RUSSIAN BELIEF THAT THE WAY YOU spend the first day of the new year determines how the rest of that year will go. I spent that morning in a sunlit New York street, wearing sandals, green pants well suited for the rain forest, and an alpaca poncho. It took two hours to dig my car out of a huge mound of snow. While we were in Ecuador, a storm dumped more snow on New York than Moscow usually gets in a whole winter.

I had three months left before my PhD defense, and Nastia had a year before hers. It was time to get serious about looking for jobs. Unfortunately, the job situation wasn't good. For decades, the American system of graduate education has been growing as a Ponzi scheme. Universities admitted more and more graduate students as a cheap workforce for teaching more and more paying undergraduates. The resulting overproduction of new PhDs made it virtually impossible to get a postdoc: for each new position, hundreds of people applied.

Things would've been easier if we'd spent our graduate school years establishing contacts with potential employers. But we were a bit naive and thought that writing an excellent dissertation was more important. I made things even worse for myself by spending so much time in the field. Now we tried to catch up by attending conferences and publishing our results. Publishing in scientific journals takes a lot of time, sometimes many years. Eventually, I had a few papers accepted in top-class journals, but that wasn't sufficient for finding employment.

There were probably ten times more positions in math than in zoology and a lot more conferences. So Nastia got so busy submitting job applications and traveling to give talks that it was putting a serious strain on her health. As for me, I had too much work finishing the thesis, so I made it to only two conferences that spring. I heard a lot of interesting stuff there. For example, a talk by researchers who attached small camcorders ("crittercams") to alligator heads and found that alligators feed underwater much more often than previously believed, snatching fish, crabs, or some other small prey every few minutes on summer nights.

As much as I was tired of living in my office and spending thirty-something hours driving to and from Knoxville every two weeks, I knew I was going to miss Miami once I graduated. Not just the weather and the wildlife, from alligators to tiny geckos in my apartment, but also all the great people in the university and the teaching, which was so much fun. Sometimes my students would drive me crazy by giving stupid answers to the simplest questions (one of them wrote, "X has low reproductive success because he has one daughter and six grandchildren, and the grandchildren aren't his children"), but generally they were smart kids and did well. Once I had a small class that was so good that I had to give an "A" grade to everyone, even though I wasn't supposed to. Among the students in that class was an Israeli guy named Ari who was seriously interested in botany. I didn't know at the time that my role in his life wasn't going to be just teaching.

A year before becoming a graduate student, I spent two months in Pakistan and Afghanistan looking for an almost-mythical animal called a woolly flying squirrel. No naturalist had ever seen it in the wild until I found that squirrel—the world's largest—one magic winter night in the snow-covered forests of Nanga Parbat.

My search often led me to remote villages where no Westerner had set foot since the time of Alexander the Great. People there were invariably fascinated by me, especially if I mentioned that I was an atheist, a creature of scary legends for many Muslims. In one village in Kashmir I met a young man named Sardar. He was a student of a shady Islamist madrassa in Peshawar, just a little bit short of a suicide-bomber school. We had a few interesting discussions on theology, often joined by the entire village. When I was about to leave, he asked me if I had any friends in Israel. He wanted to meet some Israelis on the Internet to understand what kind of people they really were.

At the time none of my Israeli friends were interested. Two years later, when I had Ari among my students, I asked him if he'd like to write to Sardar. He agreed, and soon they became friends. When I came back from Africa, Ari told me that they were planning to go to Egypt to meet in person.

More time had passed, and about a month before my defense I ran into Ari again. He was so happy to see me that at first I was taken aback a little.

"Can you imagine?" he shouted. "Sardar and I are going to Boston!"

"Why Boston?"

"We're going to get married!"

I wished them good luck. It was probably my greatest achievement as a teacher.

By March my thesis was mostly completed. I wrote a long acknowledgments section, listing almost two hundred people who had helped me with my research. Mentioning them all felt so good, like paying off old debts.

The only chapter I was still working on was the one that had given me the most trouble over the years, the overview of the "songs" of all crocodilian species. Bits of data were still trickling in from croc researchers. I also kept making new observations in zoos around Florida.

The best place to study exotic crocodilians was a little zoo with a long, unwieldy name: St. Augustine Alligator Farm Zoological Park.

It was the only place in the world to have all species of crocodilians, except for the recently split ones. Almost all researchers of crocodilian "songs" had worked there. I visited that friendly place many times and made precious rare observations of narrow-snouted, New Guinea, sacred, and Cuban crocodiles as well as false gharials.

An extra bonus I enjoyed while watching crocs at the park was a huge egret rookery on its grounds. From a boardwalk I could see dozens of egret, heron, spoonbill, and stork nests, sometimes at an arm's length. I've noted that trees growing in crocodile pools in zoos and croc farms are often chosen by wading birds for their colonies. Perhaps, just as the malimbes of Africa, egrets and herons rely on crocodiles for protection from snakes, raccoons, and other nest predators. This protection must be important enough to outweigh the loss of the few fledglings that accidentally fall in the water and get eaten.

Crocodiles, alligators, and caimans often push small twigs in front of them while swimming, especially in spring. It could be a form of play or a component of mating display. Alligators living under rookery trees in St. Augustine seemed to have small sticks on their snouts particularly often as they basked or floated. I remembered that I'd seen mugger crocodiles in Madras Crocodile Bank in India doing the same thing. Both places had huge egret rookeries. Could it be that the predators were trying to lure birds looking for nest-building material? If proven, it would be the first known case of tool use in crocodilians. I promised myself to get back to this when I have a chance.

One species I observed in St. Augustine was the Siamese crocodile. It's closely related to the saltwater crocodile but is smaller and lives in freshwater. The two species hybridize easily, producing fast-growing hybrids that are raised in huge numbers on commercial croc farms in Asia. The largest crocodile currently in captivity in the US is a saltwater-Siamese hybrid named Utan that lives in a crocodile zoo in South Carolina. He's twenty-one feet long, lazy, seldom moves, and never "sings."

The Siamese crocodile used to be common in Southeast Asia, but it has been hunted almost to extinction. Tiny populations cling to survival in the remotest corners of Cambodia, Laos, and Indonesian Kalimantan. A few crocs have been reintroduced to national parks in Thailand and Vietnam.

Utan, a saltwater-Siamese hybrid

The Siamese pair living in St. Augustine was famous for its parenting skills. It has been rumored for years that crocodiles sometimes feed their young, but such accounts were considered fairy tales by skeptics. In 2007, John Brueggen, the director of the park, videotaped the Siamese female bringing meat to her yearling kids and holding it while they tore off small pieces. This discovery has implications for our understanding of dinosaur biology: phylogenetic bracketing suggests that even the earliest dinosaurs could have had complex parental care.

John also documented the male's involvement in rearing the young—something that is now considered commonplace by croc researchers but was a big surprise at the time. Yet another of his observations was that crocs, gators, and caimans can regularly eat fruit and will even pull it off low branches. (Later, other researchers found that many crocodilians are important seed dispersers for some tree species.)

The Siamese male at St. Augustine "sang" beautifully. In a few days I recorded twelve roars and five headslaps, just what I'd expect from a habitat generalist. I finished observing him only two days before my defense.

For the last time, I drove back to Miami, passing familiar landmarks, among them Mount Trashmore, the giant landfill that is the unofficial highest point of the Florida Peninsula, and the turnoff to Cape Canaveral, where the last shuttle launches were being planned at that time. Watching those launches had been such a fantastic experience over the years. And the ugly skyscrapers of downtown Miami, the part of Florida I liked the least.

My defense was on April 1. April Fools' Day is taken very seriously in Russia, with some practical jokes being rather elaborate. While at the university, I kept the tradition alive by pulling jokes on my colleagues every year. The first time happened to be after two hurricanes hit Florida, so I sent out an email telling everybody that I'd be out of town for a week because of the new hurricane approaching. That caused a little panic. Another time I claimed that I'd found an alligator in the Everglades that could utter the word "fish" to passing fishermen. Yet another April 1 email was a fake quote from a study allegedly proving that the reason for slow driving on Miami streets was neither the average age of local residents nor the widespread use of cell phones, but an accidental introduction of tsetse flies carrying the sleeping sickness. Amazingly, some people would always buy those jokes. On the morning of my defense I sent out an email:

> Dear All,
>
> Please be careful walking around the department. As a gift for my defense, my African friends have sent me a live Nile crocodile. I was planning to donate it to Zoo Miami, but it escaped and is hiding somewhere in the building. It's not dangerous because it's only ten feet long and was fed a chimp carcass less than four months ago. If you find it, hold it by the tail and give me a call.

Believe it or not, I got six angry emails demanding that I remove the croc immediately.

On the day before the defense, my advisor, Steve, invited me to his house. That was a rare honor. I knew that most of his students had never been granted it. On that day I learned some surprising things about Steve: he was a descendant of Leon Trotsky, he had been arrested in more countries than I had, and he was the only producer of organic lychees in the lower forty-eight states.

The defense was surprisingly easy. My talk was well received, thanks in part to some nice photos I had taken over the years. On paper the thesis looked much less entertaining. It was just over a hundred pages long; three thousand hours of observations had gone into it, and my P-values were often less than 0.001. But what I was most proud of was finding answers to most of the questions I had asked myself five years earlier. Some of these answers were a bit oversimplified, and some were informed guesses in need of further testing. Future researchers will elaborate on them, find exciting new things, and probably prove me wrong in a few aspects. That's the nature of science: you virtually never get the absolute truth; there's always more work to do, more things to learn. But that's what makes science so exciting, so interesting, and so real.

I submitted all the required paperwork in advance, so that nothing would hold me up in Miami for an extra day. As soon as the defense and the banquet were over, I loaded all my stuff in the car and hit the road.

Nastia and I had planned a wonderful summer, a round-the-world trip. We were going to spend May in Italy, meet our relatives in Ukraine and Russia, go to Vietnam, where I had been invited by the World Wildlife Fund to help search for a mysterious animal called the saola, and then travel around eastern Australia. I expected finally to see Siamese and Johnston's crocodiles in the wild, not for study, just for fun. I didn't feel like doing any more croc research—not for a while anyway.

But the best part was being with Nastia all the time, not just two weekends a month. She was my best reward for all those years. And every minute the distance between us was getting shorter.

Epilogue

The best team catches the largest fish.
—Cajun proverb

I SPENT A YEAR IN KNOXVILLE, STUDYING SNAKE AND FISH behavior. After Nastia got her PhD, we got incredibly lucky and both found excellent jobs at Louisiana State University. I now have to commute a lot to get to remote field sites, but at least we are together most of the time.

The process of publishing my thesis and various observations is almost over, but there remain a few loose ends worth tying, so I try to observe crocodilians whenever I have a chance. One thing I thought I'd never be able to publish was that pig hunt by saltwater crocodiles I saw in New Guinea, with the larger croc chasing the pig into an ambush set up by two smaller ones. But one night in the swamps of the Pearl River of eastern Louisiana I observed a similar hunt by seven alligators, with the larger gators repeatedly chasing catfishes into shallows, and the smaller ones snatching them there. By that time networking possibilities for croc researchers had improved: a request I put out on Facebook resulted in a number of similar (sometimes almost identical) observations sent to me from Georgia, Australia, Venezuela, and Botswana. Finally, I could make the case that crocodilians can coordinate their moves with each other when they hunt, and they can play different roles.

Another mystery I was able to solve was about crocodiles and alligators living in ponds with egret rookeries. Do they really use small sticks as lures to catch birds looking for nesting material? As I've mentioned earlier, John Brueggen, the director of the Alligator Farm Zoological Park in St. Augustine, was well aware of this behavior and had observed successful hunts a few times. In Louisiana I also found a few places where egrets nested in alligator ponds. Their nesting season here was shorter and better defined than in Florida, and this gave me an opportunity to check if alligators floated with sticks on their snouts more often during the birds' nesting season. Indeed, there was a sharp peak in stick-displaying (as I termed this behavior) in late March and April, when the egrets were desperately looking for small sticks and twigs. But I almost never observed it in other months or in places more than a few hundred yards from a rookery. So it wasn't a coincidence; they really did use the sticks to attract birds.

This was a really cool discovery. Only capuchin monkeys and a few bird species had previously been known to use objects as lures. But alligators didn't just use lures; they also timed their behavior to their prey's seasonal activity. It looks like they are second only to humans in their ability to use advanced hunting techniques.

Our understanding of the complexity of crocodilian behavior has changed dramatically in the last forty years, and this little revolution is not over yet. But I often think that every other animal currently considered stupid and boring has its own amazing secrets. It's just that nobody has been able to discover them yet.

Meanwhile, a few crews have tried to film alligator "dances," but it has proved difficult, in part because Florida had two unusually rainy spring seasons and alligators were widely dispersed. So far, only one team has succeeded.

I was invited to come to the Everglades and help one such team film a documentary about alligators. They had really cool infrared equipment. Interestingly, alligators look black in the infrared even after basking in the sun all day. Apparently their scutes somehow channel the sun's energy inward and don't allow it to escape. Crocodiles don't have to cope with cold weather so often, and they probably lack this feature, judging by the fact that they don't look black in the infrared.

I suggested filming at Anhinga Trail, one of my first study sites. The team asked me if it would be safe to film the show's host, a young smooth-talking guy, as he would float among alligators in a small inflatable boat. I said it would, but they didn't seem convinced. I remembered how nervous I was when I used my kayak for the first time, and laughed. We began to inflate the boat. Suddenly, a huge alligator, the largest in the pond, got out of the water and walked toward us to investigate the suspicious activity in his domain. We had to retreat and wait until he calmed down. Now the idea didn't seem safe at all, but, to the host's credit, he still agreed to try. We decided that I'd float in a canoe nearby to help him if something went wrong. The host spent an hour slowly paddling through black water covered with splashing gators. It all looked beautiful in color and very spooky in infrared. He was exhilarated. To me it felt a bit weird, as if I were watching myself seven years earlier, and filming an episode from my own life, played out by someone else.

Unfortunately, I rarely have time for alligators nowadays. My new job keeps me very busy. Our team is reintroducing whooping cranes to Louisiana, which is much more interesting than it may sound. The fascinating story of our effort should someday make a fine book.